Environmental Sustainability Education for a Changing World

Erika Pénzesné Kónya • Martin Haigh
Josef Křeček
Editors

Environmental Sustainability Education for a Changing World

Springer

Editors
Erika Pénzesné Kónya
Department of Botany and Plant Physiology
Eszterházy Károly University
Eger, Hungary

Martin Haigh
Department of Social Sciences (Geography)
Oxford Brookes University
Oxford, UK

Josef Křeček
Department of Hydrology
Czech Technical University in Prague
Prague, Czech Republic

Jointly published with Capital Publishing Company, New Delhi, India

The print edition is not for sale in Afghanistan, Bangladesh, Bhutan, India, the Maldives, Nepal, Pakistan and Sri Lanka. Customers from Afghanistan, Bangladesh, Bhutan, India, the Maldives, Nepal, Pakistan and Sri Lanka please order the print book from: Capital Publishing Company, 7/28, Mahaveer Street, Ansari Road, Daryaganj, New Delhi 110 002, India.

ISBN 978-3-030-66383-4 ISBN 978-3-030-66384-1 (eBook)
https://doi.org/10.1007/978-3-030-66384-1

This Springer imprint is published by the registered company Springer Nature Switzerland AG
The registered company address is: Gewerbestrasse 11, 6330 Cham, Switzerland

Preface

For years, the mantra of the applied environmental sciences, not least those linked to the management of forests, water bodies, highland, and mountainous as well as headwater regions and the conservation of biodiversity, has been 'education, education, education'! This publication goes a step further. This is because environmental management at the landscape scale can only be sustained with the blessing and cooperation of the communities that inhabit and use these lands. The problem is that these communities need to be a part of land use policy formulation and decision-making; however, to be effective in this role, they need to know enough to be able to make informed decisions about what is best and why. This can be achieved through direct engagement with the community, which is best the option but also challenging to achieve, it can also be done through education of university students, but this process is affected by the length of time it takes people to achieve decision-making levels in their different careers. It can also be achieved through primary education, because of parents' sentimental attachment to the ideas and projects of their children, which is so often different from their relationship with their beloved but independent minded and uncommunicative teenagers.

This volume contributes to environmental education for sustainability by exploring case studies of the strategies and practices currently being developed to engage communities and the citizenry with the problems and potentials for managing their own environment for the welfare of all. It begins with an inspirational history of the several decade-long works of the Uttarakhand Environmental Education Centre (USNPSS) in Almora District of India's Lesser Himalaya. It examines the beliefs and attitudes of those who become extreme long-term volunteers on environmental sustainability projects, using a case study from Wales, and it examines the scientific outputs produced by citizen scientists working in the forests of Bohemia. It considers two Hungarian projects, one that aims to sensitise communities to their food-based ecological footprint and, the other, how to use story telling as a means to rethink the environmental future.

In higher education, it explores ways of embedding education for sustainability at the curriculum level, the role of circumstance and culture in the development of the environmental attitudes and priorities of university students, and how vocational higher education can help build sustainability, environmental sensitivity, and local community development thoughts into graduates beginning careers in Turkey's tourism industry. The reform of higher education is also a theme adopted by a chapter that considers its role in achieving a more sustainable development of the Arctic frontier in Greenland. Several further contributions consider on-going work in community and higher education for the protection of biodiversity, notably in Ukraine, but also in Bulgaria, Jordan and Pakistan, and the ways in which such concerns can be communicated through the curriculum, not least in Vietnam.

This book's subtext is to work towards the achievement of the United Nations Sustainability Goals and Targets. However, throughout, these contributions aim to show both what may be accomplished and to suggest how positive changes may be achieved, despite the many obstacles to be overcome. The aim is to inform and inspire the practitioners of environmental education for a more sustainable future.

Eger, Hungary — Erika Pénzesné Kónya

Oxford, UK — Martin Haigh

Prague, Czech Republic — Josef Křeček

Acknowledgements

Thanks go to the following colleagues who assisted the peer review processes for this book. They are:

- **G.M. Arruda**, Coventry University, London, UK
- **Jason Box**, Geological Survey of Denmark and Greenland (GEUS), Copenhagen, Denmark
- **János Mika**, Eszterházy Károly University, Eger, Hungary
- **Sumara Khan**, Coventry University, Coventry, UK
- **Natalia Rocha Lawton**, Coventry University, Coventry, UK
- **Servel Miller**, University of Chester, UK
- **Eva Pažourková**, Czech Technical University in Prague
- **AttilaVarga**, Hungarian Institute for Educational Research and Development, Eszterházy Károly University, Eger, Hungary

Contents

Contributors

Muhamad Aasim Department of Biotechnology, Faculty of Science, Necmettin Erbakan University, Konya, Turkey

Zoltán P. Alföldi Department of Plant Sciences and Biotechnology, Georgikon Faculty, University of Pannonia, Keszthely, Hungary

Ibrahim Al Rawashdeh Department of Biological Sciences, Al Hussein Bin Talal University, Ma'an, Jordan

Abdel Rahman Al-Tawaha Department of Biological Sciences, Al Hussein Bin Talal University, Ma'an, Jordan

Gisele M. Arruda Global Energy and Sustainability Programme, Coventry University, London, UK

Olesya Bezsmertna Department of Botany (and Oleksii Vasyliuk's Lab), Taras Shevchenko National University of Kyiv, Kyiv, Ukraine

Emine Cihangir Department of Tourism Management, School of Tourism and Hotel Management, Van Yuzuncu Yil University, Van, Turkey

Sarolta M. Darvay Faculty of Primary and Pre-school Education, EötvösLoránd University, Budapest, Hungary
Faculty of Education, Department of Biology, J. Selye University, Komarom, Slovakia

Martin Haigh Department of Social Sciences (Geography), Oxford Brookes University, Oxford, UK

Katalin Hill Faculty of Primary and Pre-school Education, Eötvös Loránd University, Budapest, Hungary

Anh Tuan Hoang Faculty of Mechanical Engineering, Ho Chi Minh City University of Transport, Ho Chi Minh City, Vietnam

Marianne P. Kilmartin Educational Advice and Guidance Department, Open University in the South, Oxford, UK

Josef Křeček Department of Hydrology, Czech Technical University in Prague, Prague, Czech Republic

Mariyana Lyubenova Faculty of Biology, St. Kliment Ohridski University of Sofia, Sofia, Bulgaria

János Mika Doctoral School of Education, Eszterházy Károly University, Eger, Hungary

Andacs Noémi Tudatos Vásárlók Egyesülete (Association of Conscious Consumers Hungary), Budapest, Hungary

Anuradhe Pande Uttarakhand Seva Nidhi Paryavaran Shiksha Sansthan (USNPSS; 1), Uttarakhand, India

Lalit Pande Uttarakhand Seva Nidhi Paryavaran Shiksha Sansthan (USNPSS; 1), Uttarakhand, India

Eva Pažourková Department of Hydrology, Czech Technical University in Prague, Prague, Czech Republic

Erika Pénzesné Kónya Department of Botany and Plant Physiology, Eszterházy Károly University, Eger, Hungary

Mykyta Peregrym Eszterházy Károly University, Eger, Hungary
NGO "Ukrainian Nature Conservation Group", Kyiv, Ukraine

Ali M. Qaisi Department of Pharmaceutical Sciences, School of Pharmacy, University of Jordan, Amman, Jordan

Gabriella Rózsa Eötvös Lóránd University, Budapest, Hungary

Mehmet Şeremet Department of Transportation and Business Engineering Management, Maritime Faculty, Van Yuzuncu Yil University, Van, Turkey

Attila Varga Hungarian Institute for Educational Research and Development, Eszterházy Károly University, Eger, Hungary

Oleksij Vasyliuk NGO "Ukrainian Nature Conservation Group", Kyiv, Ukraine

Van Viet Pham Department of Science, Technology and Research Development, Vietnam Maritime University, Haiphong, Vietnam

Thi Lan Anh Vu Ho Chi Minh City University of Transport, Ho Chi Minh City, Vietnam

Kamelia Yotovska Faculty of Biology, St. Kliment Ohridski University of Sofia, Sofia, Bulgaria

About the Editors

Erika Pénzesné Kónya PhD, is an associate professor of biology and head of the Department of Botany and Plant Physiology at Eszterházy Károly University, Hungary, and editor of the journal *Acta Biologica Plantarum Agriensis* (ABPA). In 2017, she was selected as vice chair of the Planta Europa Network.

In 2003 and 2007 she was awarded the Zólyominé Barna Piroska prize of the Ecological Committee at the Hungarian Academy of Sciences, and in 2007, Dr Kónya received the Science Award of Miskolc Committee at the Hungarian Academy of Sciences. Since 2015 she is the leader of the Botanical Garden of Eszterházy Károly University and has developed a school garden programme and organised the international Botanical Art Festival and Week of Botany for 10 years there. Her special interests include environmental education for a sustainable future, plant conservation and plant ecology. She is a founder member of the International Environmental Conference 2016 and 2018 (IEEC). Dr Kónya is co-leader of a project supported by EU, 'Building international research environment of light pollution (EFOP 362.16-2017-000014)', in which there is a working group dealing with environmental education.

Martin Haigh PhD, NTF, SFHEA, is emeritus professor of geography in the Department of Social Sciences at Oxford Brookes University, UK, and a former editor of the *Journal of Geography in Higher Education*. In 2010, he was awarded the Royal Geographical Society's Taylor and Francis Award for 'Excellence in Geography Teaching (Higher Education)'. His special interests include environmental education for a sustainable future, the development of dharmic pedagogies, the internationalisation of university curricula, and applied environmental reconstruction, especially in Uttarakhand and Wales, where he leads a long-running volunteer- and community-based project on the reforestation and rehabilitation of former coal lands. He is a founder member of the International Association for Headwater Control and former leader of the World Association of Soil and Water Conservation (WASWAC). He was formerly 'technical advisor' to the Bhumi Project, based at the Oxford University Centre for Hindu Studies, and, previously, to the Lotus

Trust's 'Karma to Climate Change' project. This initiative, led by young people from the UK's immigrant community, attempted to promote pro-sustainablility behaviours to householders. The project was recognised as an 'outstanding' and 'unique' example of good practice by a UNESCO team in July 2011.

Josef Křeček PhD, is associate professor of hydrology and climatology in the Department of Hydraulics and Hydrology at Czech Technical University in Prague (Czech Republic). His special interests include ecohydrology, sustainable water resources, and environmental services of mountain catchments. He is experienced with several international projects abroad (Conservation of the Maharlu Lake in Irán or Water Circulation in the Yellow River Basin) and using volunteers in his field research (Earthwatch project 'Mountain Waters of Bohemia'). Dr Křeček is a founder member of the International Association for Headwater Control and former chair of the EFC/FAO Working Party on the Management of Mountain Watersheds. For these activities, he received the award 'European Hero' from Time Magazine in 2003.

Part I
Environmental Sustainability Education: This Book's Contribution

Chapter 1
Introduction: Environmental Sustainability Education for a Changing World

Erika Pénzesné Kónya and Martin Haigh

1 Introduction

Environmental education means many more things today than in previous decades. Now, because of the over consumption of natural resources and overloading, curtailment or poisoning of natural ecosystems and, despite the denials of powerful politicians, it is overwhelmingly apparent that we are facing an emergent environmental crisis of global and civilisation-threatening proportions. As each day passes, more and more frightening data about climate change, environmental degradation and mass extinctions of key species emerge. Almost 200 species of vertebrates have gone extinct in the last 100 years (Ceballos et al. 2017). In the USA and the UK, many bird populations are in rapid decline (Rosenberg et al. 2019). Records from Europe, Asia and USA chronicle a mass extinction of insects. Hence, agriculture is also approaching a crisis, the loss of pollenisers adding to the problems of soil, water and land degradation and contamination. It is recognised that fossil fuel-based agricultural technologies are not solutions any more than are pesticide-protected monocultures; rather they are part of a larger problem (Woods et al. 2010).

Technological change continues at rates that are much faster than either ecosystems can adapt or even scientific investigation can measure its impacts. For example: the appearance of the blue diodes that aid the construction of white LED lamps has created new ecological problems (Peregrym et al. 2018). Already, this technology is distributed worldwide but little is known about the impacts of these different types of LED lights. In recent years, new research is appearing and highlighting the risk of blue and white LED lights as part of artificial light at night (ALAN) pollution

E. Pénzesné Kónya (✉)
Department of Botany and Plant Physiology, Eszterházy Károly University, Eger, Hungary
e-mail: konya.erika@uni-eszterhazy.hu

M. Haigh
Department of Social Sciences (Geography), Oxford Brookes University, Oxford, UK
e-mail: mhaigh@brookes.ac.uk

E. Pénzesné Kónya et al. (eds.), *Environmental Sustainability Education for a Changing World*, https://doi.org/10.1007/978-3-030-66384-1_1

(Knop et al. 2017; Grubisic et al. 2018). Meanwhile, global temperatures rise (Fig. 1.1), water resources are at risk, glaciers retreat, and arctic ice diminishes (Fig. 1.2) and new records are set year by year. As a consequence, wild-fires become more and more frequent and destructive and the number of extreme weather events increases. Sadly, these revelations are not new. Their emergence has been charted by the scientific literature for decades (Solomon et al. 2009). However, the new realities have created an unprecedented urgency in environmental education and a shift in its focus from informing and appreciation to fostering pro-environmental action and Education for Sustainability or rather education for survival (Kopnina 2012).

Unfortunately, environmental education has proved inadequate to the massive task of getting the message of these problems through to public consciousness and, despite some progress, a huge amount needs to be done to secure a sustainable future. Like all education, too much Environmental Education has been locked into the secondary and primary school classroom, and despite some wonderful initiatives (e.g. Waite et al. 2016; see also: https://jates.org/index.php/jatespath/issue/view/9), too little has reached out into the wider community.

Education has also a crisis caused by societal changes that have outpaced its methodological development. Generations X, Y and Z access all the information they want about the world through the internet. Much of this information is of suspect quality and veracity; they alone act as its peer reviewers. The internet brings a lot of information into learner's minds but often they lack the means to evaluate its value and how to make good decisions for their community on the basis of that information. Additionally the fact is that, for many, the virtual world is (almost) as important as their everyday life in the real, natural or urban world. Their detachment from the realities of the natural world and potential isolation from their local community is more extreme than it has ever been. For many, the natural world is outside their everyday lived experience. Meanwhile, education is, currently, struggling to undertake a digital transformation (Petkovics 2018). Similarly, Environmental Education is struggling to adapt, although some of its adaptations risk making the

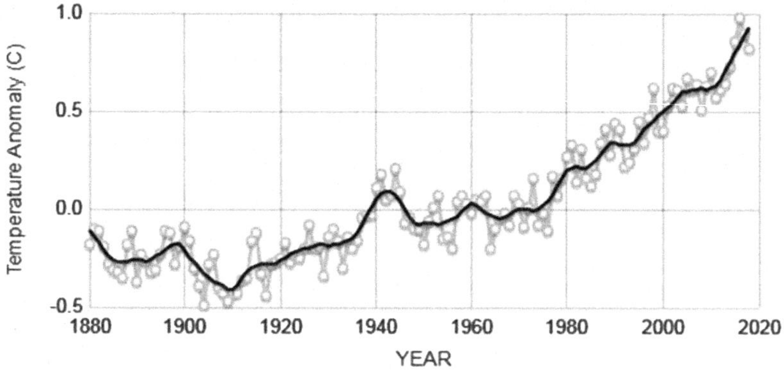

Fig. 1.1 Global temperature rise. (NASA report: NASA's Goddard Institute for Space Studies (GISS))

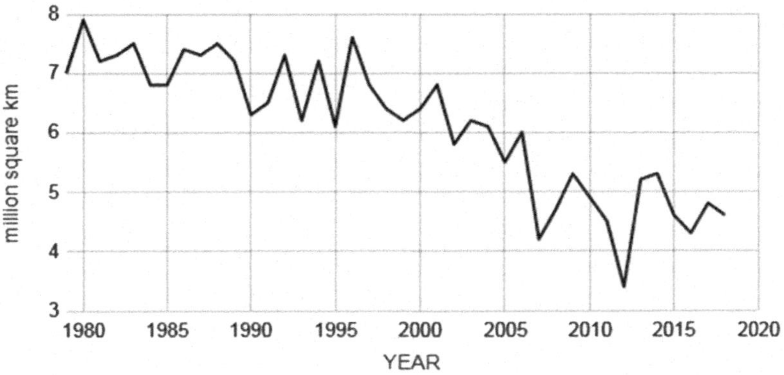

Fig. 1.2 Global ice minimum from 1980 to 2020. (*Source*: NASA report NASA's Goddard Institute for Space Studies (GISS))

problem of detachment worse; for example the Virtual Fieldtrip, which confirms the view that the world is best understood through its representation in cyberspace (Cliffe 2017).

Given the pace of technological and societal change, what is required is a more adaptive style of environmental education, something that connects the learner with the realities of their environment and encourages them to influence thinking and encourage pro-environmental behaviour in their community (Haigh 2017). Inevitably, this must be a lifelong learning process rather than just a matter of formal schooling, not least because of the dynamic nature of the environmental problems that are to be faced and the fact that engaging with the new media cannot be avoided (Benczik 2019).

2 This Book: Structure, Sources and Content

This book advances a future-oriented vision of the development of environmental and sustainability education in settings outside formal schooling and outside the formal Secondary and Primary school curriculum. It aims to provide experience-based guidance and practical advice for teacher practitioners by teaching practitioners across the broad field of environmental and sustainability education. Especially, this includes those who work in the community, outside the classroom, often with volunteers or citizen scientists, and sometimes in museums, as well as those who analyse Higher Education policy. Towards this end, some chapters explore the development of Education for Sustainable Development (ESD) in Higher Education, including its wider institutional contexts, while others explore the character, attitudes and motivations of those engaging voluntarily with environment and sustainability issues. The book is divided into three sections, the first section concerns scene setting and overview, the next deals directly with field studies of community education and engagement through environmental volunteering, followed by chapters concerning biodiversity and sustainability education. The final section deals with macroscale issues of curriculum design, policy and practice through Higher Education and National Level case studies.

The foundation stone for this collection are papers presented to the *Second International Environmental Education Conference*, held in Eger, Hungary, May 30–June 1, 2018 as part of European Sustainable Development Week (see: https://www.esdw.eu; see also IEEC papers in:https://jates.org/index.php/jatespath/issue/view/1). The conference organisers proposed, simply, that this should be a conference where good sustainable practices and their educational aspects could be introduced to a wider audience of practitioners (Mika and Pénzesné Kónya 2020). They sought papers that supported the United Nation's 2030 Agenda for Sustainable Development, its 17 Sustainable Development Goals as well as the challenges of Biodiversity Education and the long term consequences of present day actions. This meeting was also co-sponsored by the International Association for Headwater Control (IAHC) and the United Nations Food and Agricultural Organisation's (FAO) Working Group European Forestry Commission's Management of Mountain Watersheds, who held a special session on "Forest and water in mountain catchments: education, citizen science, and mountain partnership". This collaboration is reflected in the unusual proportion of chapters that are set in upland or remote 'headwater' regions in Europe and Asia.

This conference built upon the findings of the larger first *International Environmental Education Conference*, which was convened by Erika Pénzesné Kónya and János Mika at Eszterházy Károly University, Eger, Hungary, in 2016. This had the ambition of opening up the Hungarian discourse to new international viewpoints and of building bridges between educationists, artists and environmental researchers to strengthen the collaboration and cooperation between them for saving and protecting the environment. During the conference, there was a Botanical and Nature Art Festival and a Week of Botany (see: http://ieec2018.uni-eszterhazy.hu/).

Chapter 2 written by János Mika and Erika Pénzesné Kónya evaluates the topics and total contribution of the two *International Environmental Education Conferences*. These are evaluated in terms of their support of the 17 United Nations 'Millennium' Sustainable Development Goals (SDG) (2016–2030). These argue that the double task of Environmental Education is: (i) to emphasise and illustrate uniqueness, complexity and interactions among various components of the environment, and (ii) to call the attention to various threats of environmental degradation and the means for remediation. The chapter begins with an overview of the UN Sustainable Goals (SDG, 2016–2030), where sustainability is understood in human terms and so treats society and economy, not just as conditions for environmental sustainability but also, as equally important pillars in the "building of sustainability". Hence, among the 17 SDG Goals and 169 Targets, just 14 Goals and 34 Targets are considered directly related to the environment.

These SDG goals and targets are explored sequentially starting with those connected to primary and secondary human needs and then those directed to limiting human impacts. Finally, they tackle modern crisis areas. Mika and Pénzesné Kónya (2020) note that while most of these environment-related targets are in synergy with the needs of the environment, some, such as the aim of doubling agricultural production will certainly create additional sustainability problems.

However, when the contributions volunteered to the two IEEC Conferences are evaluated, relatively few of these SDG-goals and targets are addressed directly or even mentioned. This suggests that there remains a serious disconnect between the aspirations of the United Nations (with its affiliated inter-governmental agencies) and the grassroots concerns and lived experience of environmental educators. Mika and Pénzesné Kónya (2020) conclude that, to be effective, environmental education must do more to represent the problems facing the environment at all scales and also to highlight action-oriented and concrete mitigation opportunities in ways that encourage participatory action from both individuals and their communities.

Part 2 of the book focusses on informal environmental education in the community, especially working with volunteers. The section leads with an inspirational paper from two pioneering environmental educators of India, Anuradhe Pande and Lalit Pande (2020). Their chapter on '*Community Learning and Resilience in the Himalayas: Uttarakhand, India*' is a largely autobiographical account that reflects upon a 'continuing organic journey' across three decades' by the senior author, from her arrival as a graduate volunteer with the Uttarakhand Seva Nidhi, also known as the Uttarakhand Environmental Education Centre (USNPSS), to her current leading role.

The USNPSS is a key non-government organisation devoted to both social uplift and environmental improvement in Himalayan rural village communities. It is based in Almora, where its HQ supports a substantial library and teaching centre. More crucially, the USNPSS operates a practice-based environmental education and rural development programme in the small rural village communities of the Himalayas in north India. This chapter builds from long experience to suggest best practice and ethical guidance for those seeking to interact with such communities. It considers how to build: a deep, respectful, understanding of rural community life; good working relationships; mutual trust, integrity and belief in programme accountability. The original focus of USNPSS work was women and children's education. Interventions on behalf of environmental improvement and development were initially addressed to them. However, soon the programme expanded its focus towards its signature holistic ecological vision that includes all village residents, as well as school teachers, government officials and other non-government organizations, across multiple activities carried out in villages.

This chapter evaluates what community learning means in practice. It considers the challenges and tensions and how such work might increase the resilience of village ecosystems with village people playing key and integral roles. The authors emphasise that the resilience of a village ecosystem, including its community, is about more than just resistance to, adjusting to or recovering from stress or a shock and it is about more than simply a return to the pre-stress situation. Rather, it is about exploring and creating a new state of balance between people and their environment.

Chapter 4 written by Martin Haigh (2020), explores the impact of a major disaster on the environmental attitudes of university students in the same part of Himalayan Uttarakhand, India. Of course, effective environmental protection and management in this steep, fragile, mountain headwater requires community support and engagement. Building this, in turn, requires an understanding of the attitudes and values of the host community as well as some affective, conative, means of communicating with them in order to inspire new environmental policy and action (Haigh 2017). This paper examines environmental attitudes among undergraduate learners in the aftermath of a major disaster in the Kumaun Himalaya using the standard 'New Ecological Paradigm' questionnaire and some additional items. It compares these findings with results from elsewhere and also considers how they might be employed to enhance environmental education. However, the chapter's main conclusion is that standard questionnaires like 'New Ecological Paradigm' questionnaire, whose roots are found in the thinking of the USA in the 1970s, do not cross cultural boundaries at all well and that even a relatively poorly designed questionnaire that was deeply rooted in local culture and patterns of thought would have produced better and more useful findings. The paper concludes by thinking about what Indian cultural resources might be employed for such a project.

In Chap. 5, Marianne Kilmartin (2020) returns to the subject of environmental volunteering and evaluates the motivations, benefits and needs of long term volunteer contributors in environmental projects. The motivations of environmental volunteers are quite well documented but, what makes this study special is that, like Anuradhe Pande (Chap. 3), the volunteering has involved service sustained over very many years, in this case, for more than ten and in some cases more than 28 years. Less unusual is that the volunteered work has involved, often, challenging weather conditions and hard manual labour on some of the steeply-sloping, badly degraded, arguably contaminated wastelands produced by opencast coal-mine waste-disposal in upland South Wales. The work has involved planting and measuring trees, digging soil pits in rocky, compacted ground, and recording ecological conditions, in support of this project's 'Cradle for Nature' approach to land reclamation (Haigh 2018). In 2017, Kilmartin recorded interviews with seventeen volunteers during this project's annual field camp to determine why these long-term volunteers have remained willing to invest their spare time, energy and money in this community-based land reclamation project.

Kilmartin's results show that long-term volunteers have a greater sense of ownership of the project, a closer link with the environmental issues tackled and deeper friendships with other members of the volunteer group than those who have supported the project for 10 years or less. More specifically, the more than 10 years' volunteer group were: seven times more likely than the less than 10 years' volunteers to feel that the project was part of their identity. They were also six times more likely to feel a sense of ownership; more than twice as likely to want to continue with this Cradle for Nature project, and twice as likely to enjoy the social benefits, to feel that their contribution is needed and that they were making a positive contribution to the environment, Both categories shared similar reasons for volunteering in this environmental project but the longer term volunteers expressed almost twice as many reasons for remaining with the project than the shorter term volunteers. This said, there are three areas where the shorter term volunteers emphasise a particular motivation more than their longer term volunteer counterparts. These are: 1. value for their *curriculum vitae*, perhaps because the older, longer term, volunteers are mainly retired, so career development is no longer their concern; 2. desire to help the local community, which suggests that the longer term volunteers have relatively great intrinsic motivations; and 3. the enjoyment of working outdoors, which also becomes less novel and more challenging for older people.

Chapters 3, 4 and 5 all concern the human aspects of environmental volunteering and environmental action. By contrast, the focus of Chap. 6 is more the scientific outcome and value of work undertaken by 'volunteer citizen scientists'. Josef Křeček and Eva Pažourková (2020) describe the outcomes from a citizen-science-based project that supports the sustainability of environmental services, such as water supply, in the mountain forest catchments of northern Bohemia. Here, between 1991 and 2012, volunteer citizen scientists, marshalled by the Earthwatch Institute, participated in ground observations during field surveys of the Jizera Mountains, Czech Republic. Each year, four to five teams of four to eight lay participants per team participated in a two-week field study. Křeček and Pažourková (2020) evaluate the education and effectiveness of those teams with respect to the tasks of integrated ecological monitoring and science-based stakeholder dialogue in the restoration of forested headwater catchments. These citizen scientists proved motivated to progress their training, committed to quality data collection and, subsequently, were effective in ground data gathering.

Through their participation, the citizen science volunteers also learnt why having the status of 'Protective forests' cannot fully ensure the environmental services of these mountain catchments. In respect of water resources recharge, structured forestry practices should respect five main priorities within a catchment: protection of surface waters (buffer strips along reservoirs and streams), conservation of biodiversity (wetland-spots), soil protection (slopes over 30%), fog drip maintenance (elevations above 900 m), and evapotranspiration control. The volunteers' work helped delineate these critical zones in the basins of two important water-supply reservoirs, Josefův Důl and Souš, which both have a history of severe 'acid rain' deposition, which also means that the atmospheric loading of acid delivered to these catchments must be controlled by managing the tree canopy area available to intercept rain and precipitation from fog.

Education and eating is the focus of Chap. 7 by Rózsa et al. (2020). This asks if raising public consciousness could decrease the ecological (food) footprint' of society? Of course, agriculture is a major economic sector whose land-take, currently, is about 50% of the world's habitable area, which means that land management policies, especially those, like agriculture, which encourage off-site impacts by the release of pollutants and poisons, have very significant effects on the biological diversity and health of planet Earth. For example, the excess runoff of agricultural nutrients is one reason that the number of anoxic dead-zones in the oceans has (almost) doubled every decade since the 1960s. Of course, producing food for humankind (and our billions of pet animals) through growing crops or livestock also contributes 20–30% of global greenhouse gas emissions. This is why changing our habits related to food, what we eat, how we prepare our food, and how much we throw away, could make a big difference to our collective ecological footprint. Unfortunately, Hungarian public education, as in many other countries, restricts the main focus of food-education to health and hygiene. However, a small number of initiatives are trying to integrate environmental issues into food education. These include the work of the School Gardens Movement (e.g. Fischer et al. 2019) and some international projects (e.g. Eathink 2015.org, which aims to 'increase European youth awareness' of EU policies on global sustainable development and enhance students' and teachers' critical understanding of, and active involvement with, sustainable food production).

Here, a preliminary study by Rózsa et al. (2020) discusses engagement with the work of Ökokör-Kamra (EcoTeam–Food), a system of non-formal learning for adults supported by Tudatos Vásárlók Egyesülete (Association of Conscious Consumers) in Hungary. Project methodology is based on the work of a European Union funded international 'EcoTeams' programme coordinated by Global Action Plan International (Davidson 2010). This consists of establishing self-organizing learning groups of 5–10 people who learn together about food and environment related issues. In this preliminary study, 52 questionnaires and nine in-depth interviews were administered to Ökokör-Kamra group coaches and participants. Several negative reactions were reported but nearly all respondents gave positive feedback about the Ecoteam approach and more than three quarters reported long term changes in their approach to food waste, while two thirds reported positive support from their family and that they had shared their new ideas with friends and neighbours.

Chapter 8 by Peregrym, Vasyliuk and Bezsmertna describes public educational initiatives for the promotion of knowledge about rare plants in Ukraine. The level of knowledge of Ukrainian citizens about rare and endangered plants remains low. As a result, most are unable to name the majority of the species that merit legal protection in the country and that are listed in the Red Data Book of Ukraine. In Ukraine, the law based on the Red Data Book is the main and most effective tool in nature conservation, because the presence of a species from the Red Data Book within any natural or semi natural territory can be used as the basis for creating a protected area. Similarly, those responsible for harming individuals of a rare species or for destroying of its natural habitat can be punished. Consequently, public educational initiatives aimed at promoting of knowledge about rare plants in Ukraine have been initiated. They include: an historical analysis of Plant Red Listing; from 2014, a series of popular science books "50 rare plants", which concern the rare and unique species of every region in Ukraine, which are published for the information of the local population; local knowledge has also been promoted through special informational groups in social networks. Finally, local people have been encouraged to engage with citizen science through local data collecting projects about biodiversity in Ukraine.

The work of an international team, fronted by Abdel Rahman Al-Tawaha (Al-Tawaha et al. 2020), is the subject of Chap. 9. This deals with endangered plant species and also determination of the actual status of biological diversity and its protection in Jordan, Bulgaria and Pakistan and, less thoroughly, with ecological education to promote greater awareness of biological diversity. Of course, the flora of these four nations is rich and diverse, not least in medicinal and aromatic plants, herbs and spices. This chapter includes detailed information about the endangered aspects of their national flora, especially as represented in their national Red Book lists, as well as the need for its protection. Matters related to Environmental Education appear as something of an afterthought, although Bulgaria, at least, has legislation devoted to its promotion at all levels of formal education. In the other countries, development seems to be a much earlier stage, although their discussion emphasizes the importance of civic society and public consciousness-raising for the conservation of endangered species. However, while most subsections of this chapter display greater interest in their cataloguing of endangered species and official means of protection than in bringing these issues to the attention of the wider public, the team agrees that the main goals for environmental education should be consciousness raising and the development of ecological, ecosystems thinking. In Higher Education, special attention should be paid to the ecosystem role of biodiversity and the importance of preserving biodiversity for the stability and resilience of biospheric processes, the effective operation of ecosystem functions, and the continuing provision of ecosystem services for human communities. The team also agrees that activities outside the classroom and in non-formal education are very important for all levels of environmental education.

Part 3 of this book concerns environmental sustainability and sustainability education in the formal educational curriculum, especially at university level (Chapters 10, 11 and 13). This section launches with Gisele Arruda (2020), whose Chap. 10 deals with the development of environmental sustainability education within Greenland's Higher Education system. This study is based on 5 years of fieldwork in Greenland's scattered, remote, small urban communities. Of course, the Arctic is facing the brunt of climate change and its environmental impacts as well as rapid environmental, economic and socio-cultural changes due to its accelerating rate of development. The Arctic, as a whole, is a new frontier both for resource development and education at all levels. This chapter aims to set the development of Greenlandic education, especially Higher Education, in the contexts of its development process, its main drivers of change and dynamics. The chapter deploys insights from the grey literature along with informal interviews with relevant stakeholders to explain the dynamics of Greenlandic society and the role of economic development before considering how its Higher Education system has and should evolve in order to respond effectively to the different drivers of change. Arruda concludes that this process requires an open-minded approach to Education for Sustainable Development (ESD) that should be linked to a sustainability perspective founded in the United Nations' Sustainable Development Goals and its ESD frameworks.

Education for Sustainable Development (ESD) is also the focus of Chap. 11 by Emre Cihangir and Mehmet Şeremet (2020). However, their focus is the development of ESD in the Turkey's tourism-oriented Vocational Degree Programmes, which serve Turkey's vast tourist industry, which is thought to be worth somewhere between 6.2% and 12.1% of national GDP in 2018, depending on how the calculation is performed.

Of course, tourism and Sustainable Development are closely related, if not necessarily compatible, considerations, while sustainable tourism is a thriving research area. However, the contribution of vocationally-oriented tourism education to sustainable development is relatively un-explored, particularly in Turkish contexts. Hence, this chapter provides a pioneering critical insight into the country's approach to the United Nations Decade of Education for Sustainable Development 2005–2014 (DESD) and its following: Agenda for Sustainable Development, 2015–2030, with its 17 Sustainable Development Goals (SDG).

Cihangir and Şeremet (2020) explore and evaluate the extent to which Turkey's Higher Education sector is progressing vocational education, especially in tourism-oriented degree programmes, and the problems of dealing with inertial forces in such a large system. Importantly, it concludes with a set of proposals for future development. These include the recommendations that: greening and sustainable management issues should be addressed within all Higher Education Institutions' (HEI) curricula; Masters level education in the Sustainable Management of Environment/Hotels/Destinations should be encouraged and sustainability awareness and skills supported by CPD (Continuing Professional Development) training; engagement with local communities should be fostered to better support the local economy and so encourage safe-guarding of the environment; greater engagement from both practitioners and other disciplines should be developed; while ESD, both in theory and practice, should be embedded in every curriculum and developed as a fundamental graduate attribute of every Tourism student. The key change needed is a new mind-set that moves away from seeing Tourism programmes as isolated business-driven curricula and develops them as environmentally-aware, multi-disciplined, skills-driven programmes.

Chapter 12, which is prepared by another international team, this time fronted by Anh Tuan Hoang et al. (2020), continues the theme of integrating Environmental Protection Education in the curriculum as a means of building awareness of the need for environmental protection and sustainable living in the community. Building an understanding of Sustainable Development is of great importance in changing the perceptions and actions of younger generations and the community at large. Hence, education has a key role in raising the environmental awareness of all groups of people in society. A commitment to integrating the content of environmental education into the curricula of public and university education should be made a mandatory requirement. Environment Protection Education helps people in general (and students in particular) not only to build sustainability awareness but also to adopt the specific behaviours that contribute to producing a greener world.

Chapter 13 by Zoltán P. Alföldi (2020) describes an award winning sustainability education programme at the University of Pannonia. In this case, the goal is to educate people in an ecological-holistic way of thinking and reorient their future actions within a new environmental paradigm. This integrated sustainability education programme at the University of Pannonia, Georgikon Faculty, Keszthely, Hungary, has been operating for the last 15 years and includes constituents such as MSc courses in Human Ecology, Bioethics and Environmental Education, as well as other practical components. These programmes aim to reveal and teach intercultural wisdom, to combine relevant traditional ecological knowledge-base of local societies with modern scientific ideas and so co-create co-operative, knowledge- and ethic-based societies. These are considered to be key to surviving current environmental crises.

The final contribution, Chap. 14 by Katalin Hill and Sarolta Darvay, suggests a practical methodology for helping trainee teachers deliver environmental education, ESD, and understanding of the United Nations' Sustainable Development Goals (SDG), which is applicable at all educational levels and in many contexts, both formal and informal. Its aim is to embed ESD and the SDGs in teacher education using the method of unfinished stories and rewriting newspaper reports. The strategy has two educational aims: first is to provide these future educators with the means to develop their thinking, attitudes, and habits in ESD and second is to help them develop a methodological tool kit for delivering ESD. In their teacher training classes, these two tasks are implemented by means of project work. In one example, students were asked to write their own ending for a newspaper article dealing with sustainability issues. Using ideas from the Hans Christian Andersen Medal winning children's story writer Gianni Rodari (2001), who wrote multiple endings for many of his tales, the topic adopted was the problem of Cape Town's water supply running out. The project required the trainee teachers to: sensitise students to environmental and social problems, to collect information about students' perception, way of thinking and competences, and to consider the "unfinished story" as an educational method. However, results showed that the majority of these trainee teacher's final finished stories sought solutions to environmental and social problems that maintained rather than reduced current patterns of consumption, which confirms that much work still needs to be done to raise the consciousness of these future teachers.

3 Discussion

One of the key changes demonstrated by both this book and the contents of the IEEC Environmental Education Conference, which source many of the papers (Mika and Pénzesné Kónya 2020) is the rapidly changing mindset that is rapidly replacing the institutional and faintly academic orientation of EE (Environmental Education) with ESD (Education for Sustainable Development, also known as Education for a Sustainable Future), which is more pro-active, activist, interventionist and holistic in its approach to the management of both environment and society. The older, perhaps now archaic, EE focus on 'nature study' and the protection of rare species persists as witnessed by several chapters in the mid-sections of this book, but increasingly the emphasis has shifted toward something that is more urgent and whose focus on the environment contains greater emphasis on the regulation of human action (e.g. Pande and Pande 2020) and that is orientated to the wellbeing of both biosphere and future generations of humans (Thunberg 2019). ESD goes beyond informative 'education about the environment', its protection and conservation and emphasises more education for the sake of the environment; it looks more to the whole environmental system and the particular disruptive, ultimately self-destructive, role of humans within it and towards means of intervention. As David Orr (1994, p. 204) remarked, ESD's aim is to ensure that human minds are "made safe for a planet with a biosphere". ESD is about transforming human society from something that harms the environment to something that contributes to its wellbeing. It is about supporting the science that is needed to achieve this (e.g. Křeček and Pažourková 2020) and recognising what social changes are needed to mitigate the human burden on the Earth and persuading people to adopt the changes that are required, despite the inertial forces within society that resist such change (e.g. Cihangir and Şeremet 2020: Arruda 2020).

Many of these new emphases are signposted by the United Nations' SDG (Sustainable Development Goals) agenda (United Nations 2015). These SDGs, which "are interconnected", aim to "address the global challenges we face, including those related to poverty, inequality, climate change, environmental degradation, peace and justice" (United Nations 2015, p. 1); the sequence is noteworthy since matters environmental are bookended by, in ranked order perhaps, a very partial view on human causation. Nevertheless, these same influential SDGs are described, discussed and used as benchmarks by several chapters (Mika and Pénzesné Kónya 2020; Cihangir and Şeremet 2020: Arruda 2020; Hill and Darvay 2020, among others).

Of course, "Ultimately, the environmental crisis is a crisis of the mind. And likewise, appropriate development is appropriate development of the mind. We see, do and are what we think and what we think is shaped by our cultures, faiths and beliefs" (Palmer and Finlay 2003, p. xv), dealing with this crisis is far from being a simple matter. Here, for example, Haigh (2020) highlights some of the problems inherent cross-cultural and international comparisons of survey data of, in this case, learners' attitudes to the environment. This chapter concludes by discussing the problem that questionnaire design, in this case of the NEP (New Environmental Paradigm) questionnaire is both a product of its times and trapped within the culture, attitudes and politics of its creators, which, in this case study, make the questionnaire unsuitable for the task of determining the attitudes of future school teachers in the Himalaya.

Nevertheless, achieving sustainability in any culture requires two things: the first is information and awareness (e.g. Peregrym et al. 2020; Al-Tawaha et al. 2020). Unfortunately, while information is relatively easily provided, although the role of the Internet is ambiguous, even when the information is correct, it is also easily and often ignored or denied, especially when the information proves inconvenient. For example, the chief characteristic of our recent history climate change negotiations, according to (then) United Nations Assistant Secretary-General Olav Kjørven (2009), is 'everyone wanting to do as little as possible, while pushing for others to do as much as possible, whilst making sure that someone else pays the bill'. So, the second and most often absent precondition for achieving sustainability (and everything related to it) is conation, creating the will to act and providing the means to engage, through volunteering or changes in life style or even through teaching

(Rózsa et al. 2020; Kilmartin 2020; Křeček and Pažourková 2020; Pande and Pande 2020; Hill and Darvay 2020).

4 Conclusion

This book's introductory survey raises some general themes, which might be considered whilst reading. The first concerns the extreme dedication and hard work that some environmental volunteers and educators devote to their work, which in many cases is inspirational (Pande and Pande 2020; Kilmartin 2020). Several chapters detail the results of years, even decades of commitment to the goals of environmental protection and sustainability. The second is the struggle to communicate the necessity for environmental action to a larger community, which, quite frankly, is often short-sightedly preoccupied with matters far more trivial than the global environmental and sustainability crisis. There is little in this book to support the notion that wider society is or can be convinced that 'the house is on fire' as Greta Thunberg puts it (Thunberg 2019). Instead, the picture is of a small number of well-meaning activists, often working in isolation, struggling against the odds to make their own positive contribution in the community and in the classroom. Despite the activities of the United Nations, their struggle to do something that may make a difference, in a world that remains if not oblivious then in denial, is admirable but dwarfed by the scale of the emerging global crises.

Nevertheless, helping shift human minds toward living wisely, intelligently, in harmony with their habitat, conscious of their duty of care for the well being of all future generations and all the creatures of the global system and designing for such a future is the greatest challenge of our time. Human society flatters itself that it is 'civilised', capable of self-determination, foresight, planning, and able to see the 'big picture' and capable of rising above its selfish, short-sighted primitive 'animal' instincts. Certainly, now, when the signs of human-induced environmental decay and crisis are everywhere, is the time to demonstrate that this is the case and to take action.

References

Al-Tawaha, A.R., Lyubenova, M., Yotovska, K., Aasim, M., Sevinc, C., Qaisi, A.M., Alrawashdeh, I. and Křeček, J. (2020). Chapter 9: Awareness of biological diversity and endangered plant species and ecological education. *In:* Pénzesné Kónya, E., Haigh, M. and Křeček, J. (Eds). *Environmental Sustainability Education for a Changing World: Inspiration for and by Practitioners*. Springer, Cham, Switzerland & Capital Publishing Company, New Delhi.

Alföldi, Z.P. (2020). Chapter 13: An Integrated Model of Sustainability Education. *In:* Pénzesné Kónya, E., Haigh, M. and Křeček, J. (Eds). *Environmental Sustainability Education for a Changing World: Inspiration for and by Practitioners*. Springer, Cham, Switzerland & Capital Publishing Company, New Delhi.

Anh Tuan Hoang, A.T., Al-Tawaha, A.R., Lan Anh Vu, Pham Van Viet, Qaisi, A.M. and Křeček, J. (2020). Chapter 12: Integrating Environmental Protection Education in the curriculum: A measure to form awareness of environmental protection for the community. *In:* Pénzesné Kónya, E., Haigh, M. and Křeček, J. (Eds). *Environmental Sustainability Education for a Changing World: Inspiration for and by Practitioners.* Springer, Cham, Switzerland & Capital Publishing Company, New Delhi.

Arruda, G.M. (2020). Chapter 10: Development of environmental sustainability education within Greenland's Higher Education. *In:* Pénzesné Kónya, E., Haigh, M. and Křeček, J. (Eds). *Environmental Sustainability Education for a Changing World: Inspiration for and by Practitioners.* Springer, Cham, Switzerland & Capital Publishing Company, New Delhi.

Benczik, I. (2019). Edutainment in the Magic Tower. *Journal of Applied Technical and Educational Sciences*, 9(1), 42-57, https://doi.org/10.24368/jates.v9i1.67.

Ceballos, G., Ehrlich, P.R. and Dirzo, R. (2017). Biological annihilation via the ongoing sixth mass extinction signalled by vertebrate population losses and declines. *Proceedings of the National Academy of Sciences*, 114(30), E6089-E6096, https://doi.org/10.1073/pnas.1704949114.

Cihangir, E. and Şeremet, M. (2020). Chapter 11: Education for Sustainable Development (ESD) in Tourism-oriented vocational degree programmes in Turkey. *In:* Pénzesné Kónya, E., Haigh, M. and Křeček, J. (Eds). *Environmental Sustainability Education for a Changing World: Inspiration for and by Practitioners.* Springer, Cham, Switzerland & Capital Publishing Company, New Delhi.

Cliffe, A.D. (2017). A review of the benefits and drawbacks to virtual field guides in today's Geoscience higher education environment. *International Journal of Educational Technology in Higher Education*, 14(1), art. 28, https://educationaltechnologyjournal.springeropen.com/articles/10.1186/s41239-017-0066-x.

Davidson, S. (2010). Global Action Plans's EcoTeams Programme. *In:* Peters, M., Fudge, S. and Jackson, T. (Eds). *Low Carbon Communities. Imaginative Approaches to Combating Climate Change Locally.* Edward Elgar, Cheltenham, UK, pp. 157–177, eISBN: 978 1 84980 710 4.

Eathink2015 (2015). *About: Aims and Objectives.* Available: https://eathink2015.org/en/about/aims-and-objectives/ (accessed 17 December 2019).

Fischer, L.K., Brinkmeyer, D., Karle, S.J., Cremer, K., Huttner, E., Seebauer, M., Nowikow, U., Schütze, B., Voigt, P., Völke, S. and Kowarik, I. (2019). Biodiverse edible schools: Linking healthy food, school gardens and local urban biodiversity. *Urban Forestry & Urban Greening*, 40, 35–43, https://doi.org/10.1016/j.ufug.2018.02.015.

Grubisic, M., van Grunsven, R.H.A., Kyba, C.C.M., Manfrin, A. and Hölker, F. (2018). Insect declines and agroecosystems: Does light pollution matter? *Annals of Applied Biology*, 173(2), 180–189, https://doi.org/10.1111/aab.12440.

Haigh, M. (2020). Chapter 4: Influence of a disaster on the environmental attitudes of university students in Uttarakhand, India. *In:* Pénzesné Kónya, E., Haigh, M. and Křeček, J. (Eds). *Environmental Sustainability Education for a Changing World: Inspiration for and by Practitioners.* Springer, Cham, Switzerland & Capital Publishing Company, New Delhi.

Haigh, M. (2018). Chapter 32: Building a 'Cradle for Nature': A paradigm for environmental reconstruction. *In:* Prasad, M.N.V., de Campos, P.J. and Maiti, S.K. (Eds), *Bio-Geotechnologies for Mine Site Rehabilitation.* Elsevier, Amsterdam, pp. 593–616, https://doi.org/10.1016/B978-0-12-812986-9.00032-4.

Haigh, M. (2017). Connective practices in Sustainability Education. *Journal of Applied Technical and Educational Sciences*, 7(4), 6–30, https://doi.org/10.24368/jates.v7i4.16.

Hill, K. and Darvay, S. (2020). Chapter 14: Education for the United Nations' SDG (Sustianable Development Goals) in teacher training using the method of unfinished stories – Rewriting newspaper reports. *In:* Pénzesné Kónya, E., Haigh, M. and Křeček, J. (Eds). *Environmental Sustainability Education for a Changing World: Inspiration for and by Practitioners.* Springer, Cham, Switzerland & Capital Publishing Company, New Delhi.

Kilmartin, M.P. (2020). Chapter 5: Understanding the motivations, perceived benefits and needs of environmental projects. *In:* Pénzesné Kónya, E., Haigh, M. and Křeček, J. (Eds). *Environmental*

Sustainability Education for a Changing World: Inspiration for and by Practitioners. Springer, Cham, Switzerland & Capital Publishing Company, New Delhi.

Kjørven, O. (2009). Quoted in: Colwell, M., Finlay, V., Hilliard, A., Weldon, S. (Eds). *Many Heavens, One Earth: Faith Commitments to Protect the Living Planet*. Alliance of Religions and Conservation, Bath, UK, pp. 15–16.

Knop, E., Zoller, L., Ryser, R., Gerpe, C., Hörler, M. and Fontaine, C. (2017). Artificial light at night as a new threat to pollination. *Nature*, 548(7666), 206–209, https://doi.org/10.1038/nature23288.

Kopnina, H. (2012). Education for sustainable development (ESD): The turn away from 'environment' in environmental education? *Environmental Education Research*, 18(5), 699–717, https://doi.org/10.1080/13504622.2012.658028.

Křeček, J. and Pažourková, E. (2020). Chapter 6: Education strategy supporting environmental services of mountain forest catchments. *In:* Pénzesné Kónya, E., Haigh, M. and Křeček, J. (Eds). *Environmental Sustainability Education for a Changing World: Inspiration for and by Practitioners*. Springer, Cham, Switzerland & Capital Publishing Company, New Delhi.

Mika, J. and Pénzesné Kónya, E. (2020). Chapter 2: Environment-related targets in the UN Sustainable Development Goals (2016–2030) and their representation in the IEECD Conferences. *In:* Pénzesné Kónya, E., Haigh, M. and Křeček, J. (Eds). *Environmental Sustainability Education for a Changing World: Inspiration for and by Practitioners*. Springer, Cham, Switzerland & Capital Publishing Company, New Delhi.

Orr, D. (1994). *Earth in Mind*. Island Press, Washington, DC.

Palmer, M. and Finlay, V. (2003). *Faith in Conservation: New Approaches to Religions and the Environment*. The World Bank, Washington, DC, https://www.researchgate.net/publication/257668455_Faith_in_Conservation_New_Approaches_to_Religions_and_the_Environment_Martin_Palmer_Victoria_Finlay_The_World_Bank_Washington_DC_2003_dyoneyworldbankorg_182pp_August_2003_US_9_ISBN_0-8213-55597.

Pande, A. and Pande, L. (2020). Chapter 3: Community learning and resilience in the Himalaya, Uttarakhand, India. *In:* Pénzesné Kónya, E., Haigh, M. and Křeček, J. (Eds). *Environmental Sustainability Education for a Changing World: Inspiration for and by Practitioners*. Springer, Cham, Switzerland & Capital Publishing Company, New Delhi.

Peregrym, M., Pénzesné Kónya, E. and Vasyljuk, O. (2018). The impact of artificial light at night (ALAN) on the National Nature Parks, Biosphere and Nature Reserves of the Steppe Zone and Crimean Mountains within Ukraine. *Bulletin of the Eurasian Dry Grassland Group*, 39, 8–14, https://doi.org/10.21570/EDGG.PG.39.8-14.

Peregrym, M., Vasyliuk, O. and Bezsmertna, O. (2020). Chapter 8: Public educational initiatives for promotion of knowledge about rare plants in Ukraine. *In:* Pénzesné Kónya, E., Haigh, M. and Křeček, J. (Eds). *Environmental Sustainability Education for a Changing World: Inspiration for and by Practitioners*. Springer, Cham, Switzerland & Capital Publishing Company, New Delhi.

Petkovics, I. (2018). Digital Transformation in Higher Education. *Journal of Applied Technical and Educational Sciences*, 8(4), 77–89, https://doi.org/10.24368/jates.v8i4.55.

Rodari, G. (2001). *Grammaticadella fantasia*. Einaudi, http://www.scuolemarconi.edu.it/scuolemarconi/wp-content/uploads/2012/05/grammatica-della-fantasia-gianni-rodari-24899.pdf

Rosenberg, K.V., Dokter, A.M., Blancher, P.J., Sauer, J.R., Smith, A.C., Smith, P.A., Stanton, J.C., Panjabi, A., Helft, L., Parr, M., Marra, P.P. (2019). Decline of the North American avifauna. *Science*, 366(6461), 120–124, https://doi.org/10.1126/science.aaw1313.

Rózsa, G., Varga, A. and Noémi, A. (2020). Chapter 7: Education and meals – could awareness-raising decrease the food footprint of humankind? *In:* Pénzesné Kónya, E., Haigh, M. and Křeček, J. (Eds). *Environmental Sustainability Education for a Changing World: Inspiration for and by Practitioners*. Springer, Cham, Switzerland & Capital Publishing Company, New Delhi.

Solomon, S., Plattner, G.K., Knutti, R. and Friedlingstein, P. (2009). Irreversible climate change due to carbon dioxide emissions. *Proceedings of the National Academy of Sciences*, 106(6), 1704–1709, https://doi.org/10.1073/pnas.0812721106.

Thunberg, G. (2019). *No One is Too Small to Make a Difference*. Penguin, Harmondsworth, UK.

United Nations (2015). *About the Sustainable Development Goals*. United Nations, New York, https://www.un.org/sustainabledevelopment/sustainable-development-goals/

Waite, S., Bølling, M. and Bentsen, P. (2016). Comparing apples and pears? a conceptual framework for understanding forms of outdoor learning through comparison of English Forest Schools and Danish udeskole. *Environmental Education Research*, 22(6), 868–892, https://doi.org/10.1080/13504622.2015.1075193.

Woods, J., Williams, A., Hughes, J.K., Black, M. and Murphy, R. (2010). Energy and the food system. *Philosophical Transactions of the Royal Society B: Biological Sciences*, 365(1554), 2991–3006, https://doi.org/10.1098/rstb.2010.0172.

Chapter 2
Environment-Related Targets in the UN Sustainable Development Goals (2016–2030) and Their Representation in the IEEC Conferences

János Mika and Erika Pénzesné Kónya

1 Introduction

In the twenty-first century, amidst dramatic changes within both society and science, the litany of environmental problems has grown to crisis proportions and urgent actions are needed. Many of these involve the whole of society, hence a new role for environmental education has emerged (UN SDG Report 2016; CBD 2014; IPBES 2016). Today, Environmental Education has a double task: (i) to emphasize and illustrate the uniqueness, complexity and interactions of, between and among the various components of the environmental system and (ii) to call attention to and enable mitigating action against the many and various threats to the environment and biodiversity. The aim of this chapter is to focus on the latter and provide a comprehensive list of the environmental challenges. Its structure follows the United Nations' (UN) Sustainability Goals (SDG 2016–2030), where sustainability is understood in its wider sense that treats society and economy not only as conditions for environmental sustainability, but also as equally important pillars in the building of sustainability, so specifying additional goals for society.

From among the 17 Goals and 169 Targets, 14 Goals and 34 Targets relate directly to the environment. These environment-related targets, which are indicated by their numbers in the UN's SDG documentation, are introduced in a logical sequence that begins with the basic, primary, needs of human beings (2. Food; 3. Health; 6. Water; 7. Energy). These are followed by goals targeting human welfare and equity: (1. No poverty; 4. Access to education; 5. Gender equality; 10. Reduced

J. Mika
Doctoral School of Education, Eszterházy Károly University, Eger, Hungary
e-mail: mika.janos@uni-eszterhazy.hu

E. Pénzesné Kónya (✉)
Department of Botany and Plant Physiology, Eszterházy Károly University, Eger, Hungary
e-mail: konya.erika@uni-eszterhazy.hu

E. Pénzesné Kónya et al. (eds.), *Environmental Sustainability Education for a Changing World*, https://doi.org/10.1007/978-3-030-66384-1_2

inequalities). The focus then shifts toward the revision of the economic system, which should be both efficient and sustainable (8. Economic growth: 9. Innovative industry; 12. Responsible consumption and production; 13. Climate action) and then to landscapes in danger (11. Cities; 14. Life in water; 15. Life on land). The list ends by advocating worldwide cooperation (16. Peace and justice; 17. Partnerships).

The 34 environment-related targets are classified into four groups; "Conservation: combating degradation, maintaining biodiversity" (18 targets); "Environmental protection: reducing pollution and other loads (12 targets)"; "Enhancing risk resilience: reducing exposure and vulnerability to hazards (eight targets)"; and "Environmental awareness and education (four targets)". Some targets are related to more than one aspect, so the overall number of the aspects is 42.

The authors of this chapter are convinced that effective environmental education must be concrete both in presenting environmental problems and in exposing mitigation possibilities. Being acquainted with these objectives is useful for all young people, as long as the concepts are presented at their own level of knowledge. Hence, the second part of this chapter, considers the contribution of the two IEEC Conferences and this volume to the SDGs. Beside the main topic of the papers, which was strongly oriented towards education and consciousness-raising, a second order classification is performed.

2 The UN Sustainable Development Goals (2016–2030)

The United Nations' 17 Sustainable Development Goals (SDG) and detailed 169 targets for the 2016–2030 period (SDG 2015) use the term 'Sustainability' in its widest understanding, This set of the problems for humankind to solve is much wider than initial thoughts on sustainability considered (e.g. Goodland and Daly 1996; Kiss and Morelli 2016). The internal problems within society and the economic system are more than just pre-conditions for environmental sustainability, they are key players. Hence, the goals encompass a very wide range of environmental, societal and economic problems. The 17 goals as, originally compiled, are displayed as Table 2.1.

Since the SDGs are not arranged in any structure, Mika and Toth (2017) have suggested a classification of the 17 goals as follows, which will be used throughout this chapter (Table 2.2). Concerning these grouping, Goal 13 – Climate action is problematic. Climate change is the only environmental problem that is tackled as a separate goal in the SDGs. All the other problems, including the loss of biodiversity, changes in nitrogen cycle, ozone depletion, etc. are considered in connection with their effects on the vulnerable spheres or human health. Goal 13 could also be an individual group, but to avoid such a thin group, it is added to the efficient, sustainable economy group. Another aspect of Goal 13 is that the UN's Paris Agreement deals more completely with climate change, while this goal does not include all aspects of adaptation and mitigation.

Table 2.1 The UN Sustainable Development Goals (SDG 2015)

Goal 1. "End poverty in all its forms everywhere"
Goal 2. "End hunger, achieve food security and improved nutrition and promote sustainable agriculture"
Goal 3. "Ensure healthy lives and promote well-being for all at all ages"
Goal 4. "Ensure inclusive and equitable quality education and promote lifelong learning opportunities for all"
Goal 5. "Achieve gender equality and empower all women and girls"
Goal 6. "Ensure availability and sustainable management of water and sanitation for all"
Goal 7. "Ensure access to affordable, reliable, sustainable and modern energy for all"
Goal 8. "Promote sustained, inclusive and sustainable economic growth, full and productive employment and decent work for all"
Goal 9. "Build resilient infrastructure, promote inclusive and sustainable industrialization and foster innovation"
Goal 10. "Reduce inequality within and among countries"
Goal 11. "Make cities and human settlements inclusive, safe, resilient and sustainable"
Goal 12. "Ensure sustainable consumption and production patterns"
Goal 13. "Take urgent action to combat climate change and its impacts"
Goal 14. "Conserve and sustainably use the oceans, seas and marine resources for sustainable development"
Goal 15. "Protect, restore and promote sustainable use of terrestrial ecosystems, sustainably manage forests, combat desertification, and halt and reverse land degradation and halt biodiversity loss"
Goal 16. "Promote peaceful and inclusive societies for sustainable development, provide access to justice for all and build effective, accountable and inclusive institutions at all levels"
Goal 17. "Strengthen the means of implementation and revitalize the Global Partnership for Sustainable Development"

Table 2.2 Classification of the 17 (SDG) Goals (Mika and Toth 2017)

Primary needs of humans (2. Food, 3. Health, 6. Water and 7. Energy)
Equality between humans (1. No poverty, 4. Education, 5. Gender equality and 10. Reduced inequalities)
Efficient, sustainable economy (8. Economic growth, 9. Innovative industry, 12. Responsible consumption and production, 13. Climate action)
Landscapes in danger (11. Cities, 14. Life in water and 15. Life on land)
Worldwide cooperation (16. Peace and justice and 17. Partnerships)

Of the 169 targets, 126 targets have measurable, numeric indicators to be reached by a specific year (mostly 2030). These targets are set into sequence as number, number. The remaining 43 targets are denoted as number, letter. The targets promote awareness, institutional or financial actions to improve the issue. Examples of both kinds of targets as related to Goal 1 are:

1.1 By 2030, eradicate extreme poverty for all people everywhere, currently measured as people living on less than $1.25 a day.

1.b Create sound policy frameworks at the national, regional and international levels, based on pro-poor and gender-sensitive development strategies, to support accelerated investment in poverty eradication actions.

3 Targets Relevant to the Environment

The SDG targets listed in Tables 2.1, 2.3, 2.5, and 2.7 are taken from the original UN SDG document and the targets' descriptions are quotations (SDG 2015). However, the authors have selected only those targets that contain measurable

Table 2.3 Environment-related direct targets in the SDGs of primary human needs

2.4	By 2030, ensure sustainable food production systems and implement resilient agricultural practices that increase productivity and production, that help maintain ecosystems, that strengthen capacity for adaptation to climate change, extreme weather, drought, flooding and other disasters and that progressively improve land and soil quality
2.5	By 2020, maintain the genetic diversity of seeds, cultivated plants and farmed and domesticated animals and their related wild species, including through soundly managed and diversified seed and plant banks at the national, regional and international levels, and promote access to and fair and equitable sharing of benefits arising from the utilization of genetic resources and associated traditional knowledge, as internationally agreed
3.3	By 2030, end the epidemics of AIDS, tuberculosis, malaria and neglected tropical diseases and combat hepatitis, water-borne diseases and other communicable diseases
3.9	By 2030, substantially reduce the number of deaths and illnesses from hazardous chemicals and air, water and soil pollution and contamination
6.3	By 2030, improve water quality by reducing pollution, eliminating dumping and minimizing release of hazardous chemicals and materials, halving the proportion of untreated wastewater and substantially increasing recycling and safe reuse globally
6.6	By 2020, protect and restore water-related ecosystems, including mountains, forests, wetlands, rivers, aquifers and lakes
7.2	By 2030, increase substantially the share of renewable energy in the global energy mix

Table 2.4 Environment-related direct targets in the SDGs of primary human needs

Target	Conservation: combat degradation, maintain diversity	Environmental protection: reduce pollution and other loads	Enhance risk resilience: reduce exposure and vulnerability	Environmental awareness and education
2.4	+		+	
2.5	+			
3.3			+	
3.9		+	+	
6.3		+		
6.6	+			
7.2		+		
All	3	3	3	0

indicators (i.e. those represented by number ‘number and not those represented by number’ letter). Environment related targets are further classified according to their content. In Tables 2.4, 2.6, 2.8, and 2.9, the following classification summarises the four key aspects of the environmental concern and activity:

- A. Conservation: combat degradation, maintain diversity
- B. Environmental protection: reduce pollution and other loads
- C. Enhance risk resilience: reduce exposure and vulnerability
- D. Environmental awareness and education

3.1 *Primary Needs of Humans (Four Goals, Seven Targets)*

The first group of SDG goals counts seven targets, which belong to four goals. The goals for Food (No. 2), Health (No. 3), and Water (No. 6) are equally represented by two targets, whereas Energy (No. 7) is related to the environment through one target. The wording of the targets in Table 2.1 is original but the key ideas are set in italics.

Table 2.5 Environment related direct targets in the SDGs for equality between humans

1.5	By 2030, build the resilience of the poor and those in vulnerable situations and reduce their exposure and vulnerability to climate-related extreme events and other economic, social and environmental shocks and disasters
4.7	By 2030, ensure that all learners acquire the knowledge and skills needed to promote sustainable development, including, among others, through education for sustainable development and sustainable lifestyles, human rights, gender equality, promotion of a culture of peace and non-violence, global citizenship and appreciation of cultural diversity and of culture’s contribution to sustainable development
10.7	Facilitate orderly, safe, regular and responsible migration and mobility of people, including through the implementation of planned and well-managed migration policies

Table 2.6 Environment-related targets in the SDGs for equality between humans

Target	Conservation: combat degradation, maintain diversity	Environmental protection: reduce pollution and other loads	Enhance risk resilience: reduce exposure and vulnerability	Environmental awareness and education
1.5			+	
4.7				+
10.7	+	+	+	
All	1	1	2	1

Table 2.7 Environment-related direct targets in the goals of efficient and sustainable production

8.4	Improve progressively, through 2030, global resource efficiency in consumption and production and endeavour to decouple economic growth from environmental degradation, in accordance with the 10-Year Framework of Programmes on Sustainable Consumption and Production, with developed countries taking the lead
9.4	By 2030, upgrade infrastructure and retrofit industries to make them sustainable, with increased resource-use efficiency and greater adoption of clean and environmentally sound technologies and industrial processes, with all countries taking action in accordance with their respective capabilities
12.2	By 2030, achieve the sustainable management and efficient use of natural resources
12.4	By 2020, achieve the environmentally sound management of chemicals and all wastes throughout their life cycle, in accordance with agreed international frameworks, and significantly reduce their release to air, water and soil in order to minimize their adverse impacts on human health and the environment
12.5	By 2030, substantially reduce waste generation through prevention, reduction, recycling and reuse
12.8	By 2030, ensure that people everywhere have the relevant information and awareness for sustainable development and lifestyles in harmony with nature
13.1	Strengthen resilience and adaptive capacity to climate-related hazards and natural disasters in all countries
13.2	Integrate climate change measures into national policies, strategies and planning
13.3	Improve education, awareness-raising and human and institutional capacity on climate change mitigation, adaptation, impact reduction and early warning

Table 2.8 Environment-related direct targets in the Goals of efficient and sustainable production

Target	Conservation: combat degradation, maintain diversity	Environmental protection: reduce pollution and other loads	Enhance risk resilience: reduce exposure and vulnerability	Environmental awareness and education
8.4	+	+		
9.4		+		
12.2	+			
12.4		+		
12.5		+		
12.8				+
13.1			+	
13.2	+	+	+	+
13.3				+
All	3	5	2	3

Classification of the relevant SDG targets is provided in Table 2.2. The seven targets appear nine times in one or the other classes of problems or activities. Conservation, Environmental protection and Risk resilience are represented by three targets, whereas Awareness and education is not represented in any of the goals related to the Primary human needs SDGs.

Table 2.9 Environment-related direct targets in the Goals of the landscapes in danger

11.4	Strengthen efforts to protect and safeguard the world's cultural and natural heritage
11.5	By 2030, significantly reduce the number of deaths and the number of people affected and substantially decrease the direct economic losses relative to global gross domestic product caused by disasters, including water-related disasters, with a focus on protecting the poor and people in vulnerable situations
11.6	By 2030, reduce the adverse per capita environmental impact of cities, including by paying special attention to air quality and municipal and other waste management
14.1	By 2025, prevent and significantly reduce marine pollution of all kinds, in particular from land-based activities, including marine debris and nutrient pollution
14.2	By 2020, sustainably manage and protect marine and coastal ecosystems to avoid significant adverse impacts, including by strengthening their resilience, and take action for their restoration in order to achieve healthy and productive oceans
14.3	Minimize and address the impacts of ocean acidification, including through enhanced scientific cooperation at all levels
14.4	By 2020, effectively regulate harvesting and end overfishing, illegal, unreported and unregulated fishing and destructive fishing practices and implement science-based management plans, in order to restore fish stocks in the shortest time feasible, at least to levels that can produce maximum sustainable yield as determined by their biological characteristics
14.5	By 2020, conserve at least 10 per cent of coastal and marine areas, consistent with national and international law and based in the best available scientific information
15.1	By 2020, ensure the conservation, restoration and sustainable use of terrestrial and inland freshwater ecosystems and their services, in particular forests, wetlands, mountains and drylands, in line with obligations under international agreements
15.2	By 2020, promote the implementation of sustainable management of all types of forests, halt deforestation, restore degraded forests and substantially increase afforestation and reforestation globally
15.3	By 2030, combat desertification, restore degraded land and soil, including land affected by desertification, drought and floods, and strive to achieve a land degradation-neutral world
15.4	By 2030, ensure the conservation of mountain ecosystems, including their biodiversity, in order to enhance their capacity to provide benefits that are essential for sustainable development
15.5	Take urgent and significant action to reduce the degradation of natural habitats, halt the loss of biodiversity and, by 2020, protect and prevent the extinction of threatened species
15.7	Take urgent action to end poaching and trafficking of protected species of flora and fauna and address both demand and supply of illegal wildlife products
15.8	By 2020, introduce measures to prevent the introduction and significantly reduce the impact of invasive alien species on land and water ecosystems and control or eradicate the priority species

3.2 *Equality Between Humans (Three Goals, Three Targets)*

The second group of SDG goals includes three targets, which belong to three goals (Table 2.5). The goals for No Poverty (No. 1), Education (No. 4), and Reduced Inequalities (No. 10) are equally represented by one target, whereas in case of Gender equality (No. 5), arguably – because of the different roles played by male and female in so many cultures, there is no direct relatedness to the environment.

A classification of the relevant SDG targets is provided in Table 2.6. The three targets appear five times in one or other classes of problems or activities. Conservation, Environmental protection and Awareness and education are represented by one target each. Risk resilience is represented by two goals related to Equality between humans (Table 2.7).

3.3 Efficient, Sustainable Production (Four Goals, Nine Targets)

The third group of SDG goals counts nine targets which belong to four goals. The goals for Economic growth (No. 8) and Innovative industry (No. 9) are represented by one target, each. At the same time, the goal for Responsible consumption and production (No. 12) and Climate action (No. 13) are represented by four and by three targets, respectively.

Classification of the relevant SDG targets is provided in Table 2.8. The nine targets appear 13 times in the four classes of problems. Conservation and Education are represented by three targets. Environmental protection and Risk resilience are represented by five and two times, among the targets concerning efficient and sustainable production, respectively (Table 2.9).

3.4 Landscapes in Danger (Three Goals, 15 Targets)

The last group of SDG goals counts 15 targets which belong to three goals. The goals for the Cities (No. 11), for Life in water (No. 14) and Life on land (No. 15) are represented by three, five and seven targets, respectively.

Classification of the relevant SDG targets is provided by Table 2.10. The 15 targets appear exactly 15 times in the four classes of problems. Conservation is represented by 11 targets, Environmental protection by three and Risk resilience by one. Education is not directly represented in any target connected with the 'Landscapes in Danger' SDGs.

Strangely, the Worldwide Cooperation SDGs (Peace and Justice, No. 16. and Partnerships, No. 17) include no direct connection to environmental issues, although several general targets should also be considered and applied in the solution of environmental problems. These matters are, however, essential to the implementation of nearly all mitigation strategies.

Table 2.10 Environment-related direct targets in the SDG Goals of the landscapes in danger

Target	Conservation: combat degradation, maintain diversity	Environmental protection: reduce pollution and other loads	Enhance risk resilience: reduce exposure and vulnerability	Environmental awareness and education
11.4	+			
11.5			+	
11.6		+		
14.1		+		
14.2	+			
14.3		+		
14.4	+			
14.5	+			
15.1	+			
15.2	+			
15.3	+			
15.4	+			
15.5	+			
15.7	+			
15.8	+			
All	11	3	1	0

4 The IEEC Conferences and Their Publications

Sustainability Education has two essential components: Education for Sustainability and Education as Sustainability (Medrick 2013). Education for Sustainability concerns the means by which citizens can be educated in how to achieve global and local sustainability. Education as Sustainability is based on lifelong learning, recognition of diversity, cooperation and collaboration, personal reflection and values, integrative understanding, responsibility and faith in other people. The IEEC Conferences have attempted to address both aspects.

The idea of funding an International Conference for Environmental Education (IEEC) emerged from a meeting at Society for Ecological Restoration's 9th European Conference, held in Oulu, Finland during August, 2014, which contained a small subsection for environmental education. Respecting the need for greater exposure for the applied aspects of Environmental Education, the IEEC conferences were founded in Eger, Hungary, as a forum for the sharing of different educational practices and concepts in the context of a special, thematically open, international conference that could focus on new ideas and communicate innovations quickly into the scientific community (Pénzesné and Mika 2017).

One special focus is biodiversity education, which is strongly connected to sustainability (see Chaps. 8 and 9). This is a large but mainly invisible problem and it is said that the people of the twenty-first century have plant blindness. Plant blindness begins in childhood and is exacerbated by the scant attention paid to botanical content in schools. For example, in South Africa, only about 11 h are devoted to plant related content in the foundation phase at school (grade R-3), while in the senior phases (grade 7–9), only 11 h are devoted to content that's specifically focused on plants (Abrie 2016).

In 2010, the parties of the Convention on Biological Diversity (CBD) adopted the Strategic Plan for Biodiversity 2011–2020 with the mission of halting biodiversity loss and enhance the benefits it provides to people. The 20 Aichi Biodiversity Targets (Aichi Targets), which are included in the Strategic Plan, are organized under five Strategic Goals. They show, for example, that two key indicators of the state of biodiversity within CBD Strategic Goal C: "Improve Status of biodiversity", namely the Living Planet Index and the Red List Index, both show ongoing decline (CBD 2014). The IPBES Global Biodiversity Assessment has similar concerns (IPBES 2016). Meanwhile, new issues are emerging such as the negative impacts of artificial light at night on pollination and animal behaviour (Peregrym et al. 2018). In Europe, this is being blamed for catastrophic declines in insect numbers, this so-called 'insect apocalypse' is now headline news (Carrington 2019).

4.1 Classification of IEEC Presentations

Tables 2.9, 2.10, and 2.11 list and classify the IEEC's published contributions, namely, this book and the two thematic issues of the on-line, open access, *Journal of Applied Technical and Educational Sciences* (jATES).

The book contributes 12 chapters, not including this one. Five chapters emerge, directly, from IEEC 2018 presentations and seven from international authors or teams who were unable to attend IEEC 2016 in person, many of them associated with IEEC 2018s collaboration with the FAO/EFC Working Group on the Protection of Mountain Watersheds.

From the presentations made to IEEC 2016 (Table 2.9), many spoken presentations subsequently appeared in the thematic issues of jATES (jATES 2017/4: https://jates.org/index.php/jatespath/issue/view/1). Three further papers were contributed to this issue by those authors who did not participate in IEEC 2016, or who presented papers on topics that were not covered by their presentations to IIEC 2016 (Haigh 2017; Poudrier 2017; Nagy 2017; Mónus 2019; Mika and Toth 2017). IEEC 2018 contributed two spoken presentations and two poster presentations that were subsequently published in a second thematic issue of *jATES* (*jATES* 2019/1: https://jates.org/index.php/jatespath/issue/view/7) along with four papers from authors who could not participate in the conference in person. This book publishes a further five papers derived directly from IEEC 2018 presentations.

Table 2.11 presents the 26 papers published in the two thematic issues of jATES and this book and classifies each according to the SDG Goal classification of environment-related topic introduced in Sect. 2.3 of this chapter. Naturally, because these conferences address Environmental Education, the list is dominated by topics that address category D (Awareness and Education).

Table 2.11 Content analysis of this book and the two jATES thematic issues based on the IEEC conferences

Author(s) and title	Primary topic	Secondary topic	Talk	Paper
This book (excluding the present chapter)				
Anuradha Pande and Lalit Pande. Community Learning and Resilience in the Himalaya, Uttarakhand, India.	D	B	No	Yes
Martin Haigh. Influence of a disaster on the environmental attitudes of university students in Uttarakhand, India	D	C	Yes	Yes
Marianne P. Kilmartin. Understanding the motivations of long-term volunteers on environmental project, perceived benefits and needs	D	D	No	Yes
Josef Křeček and Eva Pažourková. Education strategy supporting environmental services of mountain forest catchments	B	D	Yes	Yes
Gabriella Rózsa, Attila Varga and Andacs Noémi. Education and meals – could awareness-raising decrease the food footprint of humankind?	D	A	Yes	Yes
Mykyta Peregrym, Oleksij Vasyliuk and Olesya Bezsmertn. Public educational initiatives for promotion of knowledge about rare plants in Ukraine	A	D	No	Yes
Abdel Rahman Al-Tawaha, Mariyana Lyubenova, Kamelia Yotovska, Muhammad Aasim, Canan Sevinc, Ali M. Qaisi, Ibrahim Alrawashdeh and Josef Křeček. Awareness of biological diversity and endangered plant species and ecological education	D	A	No	Yes
Gisele M. Arruda. Development of environmental sustainability education within Greenland's Higher Education	D	C	No	Yes
Emine Cihangir and Mehmet Şeremet. Education for Sustainable Development (ESD) in Tourism oriented Vocational Degree Programmes in Turkey	D	C	No	Yes
Anh Tuan Hoang, Abdel Rahman Al-Tawaha, Lan Anh Vu, Van Viet Pham, Ali M. Qaisi, and Josef Křeček. Integrating Environmental Protection Education in curriculum: A measure to form awareness of environmental protection for the community	D	D	No	Yes
Zoltán Péter Alföldi. An Integrated Model of Sustainability Education	D	D	Yes	Yes
Katalin Hill and Sarolta Darvay. Education for SDGs in teacher training using method of unfinished stories: rewriting newspaper reports	D	D	Yes	Yes
***Journal of Applied Technical and Educational Sciences* (jATES) 9(1), 2019**				
Réka Könczey, Katalin Czippán. Preliminary examination of potential motifs of ecological origin as identity elements in an adult group, Hungary	D	D	No	Yes
Ferenc Mónus. Comparing environmental awareness of Hungarian students in secondary schools with different socio-economical background	D	D	Yes	Yes

(continued)

Table 2.11 (continued)

Author(s) and title	Primary topic	Secondary topic	Talk	Paper
Krisztina Szalay, Károly Antal, Zsuzsanna Emri. Environmental, non-specific factors affecting legal and illegal drug consumption	C	D	No	Yes
Izabella Benczik Edutainment in the Magic Tower: Environmental awareness as a lifelong learning process	D	D	Yes	Yes
Zsuzsanna Angyal, Fanni Négyesi. Borbala Nature Trail: Recovery potential for abandoned mining and industrial areas based on a planned nature trail	D	B	No	Yes
Éva Kovács-Bokor, Endre Domokos, Endre Kiss. Effect of industrial sludge-soil mixtures on germination of white mustard and wheat	B	B	No	Yes
Lívia Kürti. Water as theme in the Hungarian educational system	D	A	Yes	Yes
Éva Nagy. Biodiversity knowledge elements in Biology education: the base of critical thinking	A	D	Yes	Yes
***Journal of Applied Technical and Educational Sciences* (jATES) 7(4) 2017**				
Martin John Haigh. Connective Practices in Sustainability Education	D	D	No	Yes
Poudrier Claude. Environmental Education and Active Citizenship	D	D	No	Yes
Tímea Taligás. Overview of main environmental challenges and the educational possibilities to tackle them	D	D	Yes	Yes
János Mika. Education in the Sustainability Development Goals (2016-2030), sustainability in the education	ABC	D	Yes	Yes
Éva Nagy. The Comparative Analysis of the Biological Diversity in Schools	A	D	Yes	Yes
Csaba Rigóczki, Andrei Damsab, Györgyi-Ambró Kristóf. Gamification on the edge of educational sciences and pedagogical methodologies	D	D	Yes	Yes
Ferenc Mónusa, Christine Lechner. An innovative way in education for sustainable development: e-School4s –e-school for sustainability in the Danube region	D	D	Yes	Yes

n.b. The heading 'primary' and 'secondary' topic corresponds to the SDG (2016–2030) classification: A – Conservation, B – Environmental protection, C – Resilience, D – Awareness and education

Table 2.12 Summary of the three publications from the IEEC initiative

Publication	Unique Sources	A	B	C	D	ABC
Book	11	1 (2)	1 (1)	0 (2)	9 (6)	
jATES (2019)	8	1 (1)	1 (2)	1 (0)	5 (5)	
jATES (2017)	7	0 (0)	0 (0)	1 (0)	5 (7)	1(0)

Table 2.12 provides a summary of Table 2.11 and its primary and secondary focuses. Not surprisingly, class D, i.e. Awareness and Education is the most common primary and secondary focus of these 26 published papers.

The numbers indicate the primary focus of the papers; those in brackets summarise secondary ones.

4.2 *Classification of the Studies Presented in the IEEC Conferences*

As for the conferences, IEEC 2016 hosted 40 presentations (Table 2.13). Education and awareness raising (D) was the major focus of 37 studies. Two studies focussed on three other aspects (A, B and C) of the environmental problems and educational aspects were given secondary importance. One study tackled vegetation biodiversity, classified as Conservation (A). In 17 cases, the secondary focus was also education and awareness raising (D), while the next most popular secondary focus Conservation (A) with 13 studies. Environmental protection (B) was the key secondary topic of nine studies. Resilience was not a secondary focus of any study.

IEEC 2018 hosted 29 oral or poster presentations. Education and awareness raising (D) was the primary focus of 24 studies and one study emphasised the three other aspects (A, B and C) of the environmental problems while giving the educational aspects secondary importance. One study emphasised biodiversity, classified as Conservation (A), while three studies emphasised Resilience (C) as their main focal topic. In 13 cases, the secondary focus was education and awareness raising (D), followed by Conservation (A) and Environmental protection (B) with six studies each. Resilience (C) was the main topic of two studies and there were two studies tackling the three topical aspects (A, B and C) together as their secondary focus.

5 Conclusion

The total number of directly environment-related targets is 34 i.e. 20% of the overall 169 targets of the SDG (2016–2030). These are attached to 14 of the 17 goals; Goals No. 9, 16 and 17 do not contain any directly environment-related target. From the four topical aspects represented in the columns of Tables 2.4, 2.6, 2.8 and 2.10, Conservation is tackled by 18 targets, whereas Pollution and other loads are represented by 12 targets. Resilience and Awareness-raising have eight and four targets, respectively. This means that environment-related targets are quite focused.

The IEEC conferences' 69 oral or poster presentations and additional papers are listed in Tables 2.9 and 2.11, 2.12, 2.13. The most frequent primary classification, of course, was Education and Awareness raising (D), which was tackled by 61 studies and was the main focus of the two IEEC Conferences. Three studies addressed the three other aspects (A, B and C) of the environmental problems and in these

Table 2.13 Content of the studies presented in the two IEEC conferences

Author(s) and title	Primary topic	Secondary topic	Talk	Paper
IEEC 2016				
Bengt-Gunnar Jonsson. A Multidisciplinary Summer University for sustainable development	D	D	Yes	No
Claude Poudrier. Citizen Engagement: an important key to our Life	D	D	Yes	Yes
Antal Radvánszky. Environmental education and promotion of conservation messages in Botanical Gardens	D	A	Yes	No
Zsolt Molnár. Nature as source of knowledge: Traditional ecological knowledge of herders and farmers and its uses in environmental education	D	A	Yes	No
Radvánszky, Antal, Vince Zsigmond. Environmental education and promotion of conservation messages in Botanical Gardens	D	A	Yes	No
János Mika, Ilona Pajtók-Tari. Systematisation of research on education of and by climate change	D	B	Yes	No
Gerhárd Golen. Birdlife Hungary in environmental education	D	A	Yes	No
Busisiwe Gude Ndawonde. Environmental education on sustainable harvesting of medicinal plants in Northern Kwa Zulu Natal, South Africa	D	B	Yes	No
Zsanett Hajdú, Varga Anna, Biró Marianna, Demeter László, Gellény Krisztina, Bódis Judit, Miókovics Eszter, Molnár Zsolt, Molnár Krisztina. Integration of Traditional Ecological Knowledge as an Innovation in Environmental Education: Concepts, Methods and Case Studies from East-Central Europe	D	A	Yes	No
Barbara Sallee-Kereszturi. Ecological Applications - The Missing Link in Sustainable Education	D	A	Yes	No
István Erőss. Nature art education at Károly Eszterházy University (Botanic Art 2016)	D	D	Yes	No
Martin Haigh. Engaging learners with environmental education through participation in a Community Reforestation Project	D	D	Yes	Yes
Ádám Kertész. Educational aspects of sustainable landscape use	D	B	Yes	No
Josef Krecek, Zuzana Hořická, Eva Pažourková. Environmental education and stakeholder dialogue in the management of headwater catchment	D	B	Yes	No
Gyula Lakatos, Imre Kárász. Education for sustainability in the Environmental BSc in Hungary	D	D	Yes	No
Evren Cabi. Diversity of wild relatives of wheat in Turkey	A	D	Yes	No

(continued)

Table 2.13 (continued)

Author(s) and title	Primary topic	Secondary topic	Talk	Paper
Evren Cabi, Engin Kabataş, Ekrem Kurt. The Ornamental plants of Namık Kemal University (Tekirdağ, Turkey): Contribution to Educational practices	D	D	Yes	No
András Halbritter. School gardens' renaissance: How to proceed through some obstructive circumstances?	D	A	Yes	No
László Lakatos, János Mika, András Rázsi: Environmental education via subjects on atmospheric sciences	D	B	Yes	No
Mária Angelika Tóth. Examination of environmental attitudes at the Affiliated School of Eszterházy Károly University	D	D	Yes	No
Zoltán Murányi, Marianna Farkasné Ökrös. Mobilized chemistry by means of m-learning for green chemistry	D	D	Yes	No
Ildikó Katona, Emese Szitta. Primary and Pre-school Education Students about Environmental Education	D	D	Yes	No
Éva Nagy, Erika Pénzes-Kónya. Biodiversity as the tool of environmental education	D	A	Yes	Yes
Gabriella Leskó, Éva Nagy. Opportunities of Forest Pedagogy in an open-air school	D	A	Yes	No
Ibolya Revákné Markóczi, Edina Malmos, Erzsébet Jász, Nagy Miklósné Csákberényi, Enikő Kovács, Judit Ütőné Visi. Investigation of concepts related to energy culture using the word association method at primary level	D	B	Yes	No
Ádám Bohár. Necessity and possibilities of environmental education in the case of Róka-hegy Nature Conservation Area	D	A	Yes	No
András Rázsi, Boglárka Tóth, János Mika. Our virtual teachers: Mass media on climate change in Hungary	D	B	Poster	No
Cseh Gáborné Nagy Emőke. The environmental educational impact on the viewed and seen world from Antiquity to the Renaissance	D	D	Poster	No
Bíró M., Halácsy Á., Kecskés F., Molnár Zs. Naturalness Measurer for Everybody – A Citizen Science Tool for Environmental Education	D	B	Poster	No
András Schmotzer, Jana Táborská. Enhancement of Public Participation in Invasive Plant Management in the Middle Ipoly Floodplain Area (NE Hungary)	D	A	Poster	No
Andrea Sass-Gyarmati, Erika Pénzesné Kónya, Jana Táborská. Connecting Biodiversity Research and Collections with Environmental Education at Eszterházy Károly University Department of Botany and Ecology	D	A	Poster	No
Jana Táborská, Erika Pénzesné Kónya. Educational opportunities within the Eszterházy Károly University Botanical Garden	D	A	Poster	No
Richárd Novák. Astronomical education in a forest school	D	D	Poster	No

(continued)

Table 2.13 (continued)

Author(s) and title	Primary topic	Secondary topic	Talk	Paper
Piroska Tóth. Environmental Measurements, Data Processing and Evaluation according to the GLOBE Programme	D	B	Poster	No
Zsuzsanna Angyal, Gábor Ballabás. Courses outside the university: environmental education in two degree programs of the Eötvös Loránd University	D	D	Poster	No
IEEC 2018				
Ilona Pajtókné Tari. Thought-provoking strategies for teaching sustainability and environmental risks	D	ABC	Yes	No
Hideaki Marui. Environmental Education for Disaster Mitigation in Mountain Watersheds	D	C	Yes	No
Eero Kubin. Field courses for environmental education - experiences from the boreal forest environment	D	A	Yes	No
Ádám Kertész. The role of conservation agriculture in combatting land degradation. Education of farmers and decision makers	B	D	Yes	No
Eva Pazourkova:Josef Krecek. Citizen science for restoration of mountain catchments from acidification and commercial forestry practices	D	B	Yes	No
Edward Pierzgalski. Educational aspects of the international interreg project – water management in Baltic forests (WAMBAF)	D	B	Yes	No
Stanimir Kostadinov. University education in Serbia to combat soil and water degradation	D	B	No	No
Miklós Matolcsy. Sustainability Theme Week in *the Hungarian public education*	D	D	Yes	No
András Tenk, Ferenc Kecskés. The role of the Budapest's nature conservation areas in the environmental education	D	A	Yes	No
Tímea Kollarics. International comparative research of experience trails	D	A	Yes	No
Attiláné Mikó. Teaching sustainability in primary classes	D	D	Yes	No
János Mika. Recent tendencies of the environmental indicators in Hungary with applications for education	B	D	Yes	No
Gabriella Földes –Leskó. The role of Forest pedagogy in forest schools	D	D	Yes	No
László Lakatos, János Mika. Environmental prospects for 2048 as a challenge for education	ABC	D	Poster	No
Zoltán Tóth. How can a simple nature study and conservation camp serve different purposes?	D	A	Poster	No
Krisztina Szalay. Leader environmental effects on drug use	C	D	Poster	Yes
Izabella Benczik. Edutainment in the Magic Tower: Environmental awareness as a lifelong learning process	D	D	Poster	Yes
Mykyta Peregrym. An educational initiative for promotion of knowledge about rare plants in Ukraine	D	A	Poster	No

(continued)

Table 2.13 (continued)

Author(s) and title	Primary topic	Secondary topic	Talk	Paper
Erika Pénzesné Kónya, Richárd Novák. Education in Dark Sky Parks	D	B	Poster	No

n.b. The heading 'primary' and 'secondary' topic corresponds to the SDG (2016–2030) classification: A – Conservation, B – Environmental protection, C – Resilience, D – Awareness and education

papers, Educational were given secondary importance. Two studies prioritised Conservation (A) while in three Resilience (C) was the main focus.

In 30 studies, Education and Awareness raising (D) was the secondary focus followed by Conservation (A) with 19 studies and Environmental Protection (B) with 15 studies. Three studies had topical aspects A, B and C together as their secondary focus while Resilience (C) was the secondary focus of just two.

If there is a problem for future IEEC meetings to consider, it is to do more to address issues beyond education and consciousness raising and do more to target practical, more specifically, matters of Environmental Protection and Conservation. Given the greater problems facing the world, those which are driving the current climate change emergency and the catastrophic consequence of extinction and biodiversity decline, it would seem self-evident that more has to be done to link education and consciousness raising with the increasing environmental and social necessity to develop resiliency and the ability to cope in a rapidly changing and increasingly unpredictably hazardous future. Hopefully, this volume takes some small steps in this direction.

References

Abrie, A.L. (2016). The botanical content in the South African curriculum: A barren desert or a thriving forest? *South African Journal of Science* 112(1/2), Art.#2015-0127, 7 pp. https://doi.org/10.17159/sajs.2016/20150127.

Carrington, D. (2019). Artificial light blamed for 'insect apocalypse'. *The Guardian* 23 November, 2019, pp. 1–2, https://www.theguardian.com/environment/2019/nov/22/light-pollution-insect-apocalypse.

Goodland, R. and Daly, H. (1996). Environmental Sustainability: Universal and non-negotiable. *Ecological Applications*, 4(6), 1002–1017.

CBD (2014). Global Biodiversity Outlook: A mid-term assessment of progress towards the implementation of the Strategic Plan for Biodiversity 2011-2020, https://www.cbd.int/gbo/gbo4/publication/gbo4-en-hr.pdf.

IPBES (2016). Summary for policymakers of the assessment report of the Intergovernmental Science-Policy Platform on Biodiversity and Ecosystem Services on pollinators, pollination and food production. *In:* S.G. Potts, V.L. Imperatriz-Fonseca, H.T. Ngo, J.C. Biesmeijer, T.D. Breeze, L.V. Dicks, L.A. Garibaldi, R. Hill, J. Settele, A.J. Vanbergen, M.A. Aizen, S.A. Cunningham, C. Eardley, B.M. Freitas, N. Gallai, P.G. Kevan, A. Kovács-Hostyánszki, P.K. Kwapong, J. Li, X. Li, D.J. Martins, G. Nates-Parra, J.S. Pettis, R. Rader and B.F. Viana (Eds). Secretariat of the Intergovernmental Science-Policy Platform on Biodiversity and Ecosystem Services (IPBES), Bonn, Germany, pp. 1–36.

Haigh, M. (2017). Connective Practices in Sustainability Education. *Journal of Applied Technical and Educational Sciences*, 7(4), 6–30, https://doi.org/10.24368/jates.v7i4.16.

Kiss, K. and Morelli, J. (2016). *Discussing Sustainability*. Lambert Academic, Riga, Latvia.

Medrick, R. (2013). A pedagogy for Sustainability Education. *Journal of Sustainability Education*, 5, 1–3, http://www.jsedimensions.org/wordpress/wp-content/uploads/2013/06/rickMedrickAPedgagogy-ForExperientialEducationPDFReady.pdf.

Mika, J. and Tóth, B. (2017). Environmental aspects of the UN Sustainable Goals (2016-2030) (in Hungarian). *In:* A TanulásÚjÚtjai, Hera Évkönyvek 2016. Mrázik, J. (Ed.). HERA, Budapest. pp. 549–569, http://hera.org.hu/wp-content/uploads/2014/02/HERA_evkonyv_kiegesz-ites20170615.pdf.

Mónus, F. (2019). Comparing environmental awareness of Hungarian students in high-schools with different socio-economical background. *Journal of Applied Technical and Educational Sciences*, 9(1), 17–27, https://doi.org/10.24368/jates.v9i1.68.

Nagy, É. (2017). Comparative analysis of the Biological Diversity in Schools. *Journal of Applied Technical and Educational Sciences*, 7(4), 62–78.

Pénzesné Kónya, E. and Mika, J. (2017). Content of the present Issue. *Journal of Applied Technical and Educational Sciences*, 7(4), 2–4.

Peregrym, M., Kónya, E.P. and Vasyliuk, O. (2018). The impact of artificial light at night (ALAN) on the National Nature Parks, Biosphere and Nature Reserves of the Steppe Zone and Crimean Mountains within Ukraine. *Palaearctic Grasslands*, 39, 8–14, https://doi.org/10.21570/EDGG.PG.39.8-14.

Poudrier, C. (2017). Environmental Education and active citizenship. *Journal of Applied Technical and Educational Sciences*, 7(4), 31–36, https://doi.org/10.24368/jates.v7i4.7.

SDG (2015). United Nations Resolution A/RES/70/1 of 25 September 2015. New York, United Nations (The Sustainable Development Goals are listed as §51), http://www.un.org/ga/search/view_doc.asp?symbol=A/RES/70/1&Lang=E.

UN SDG Report (2016). The Sustainable Development Goals Report. United Nations, New York, http://www.un.org.lb/Library/Assets/The-Sustainable-Development-Goals-Report-2016-Global.pdf.

Part II
Informal Environmental Education in the Community and Working with Volunteers

Chapter 3
Community Learning and Resilience in the Himalayas: Uttarakhand, India

Anuradhe Pande and Lalit Pande

1 Introduction

This chapter shares insights generated from a practice-based environment education programme started by Uttarakhand Environmental Education Centre (USNPSS) in Almora District in the Himalayan mountains of north India. USNPSS is a registered non-government organisation that began to work in Uttarakhand in 1987. Its programme seeks to improve the habitat and welfare of local rural communities. However, achieving this involves both building a deep understanding of rural life and its challenges and building trust in the people that the programme would wish to serve. In this, it is important to examine the ethics of these interactions with rural communities and to ensure the integrity, sustainability and accountability of the programme.

Originally, the focus was the education of women and children on issues of environment and development. However, soon, the educational activities were expanded to include the involvement of all village residents, school teachers, government officials and other non-government organisations; in fact all those engaged in multiple activities in the villages. Given the current popularity of the term "community learning", it is tempting to portray this work as an illustration of how rural communities learn, evolve and change over time. This chapter focuses on describing what community learning means in practice and on some of its challenges and tensions. However, reflecting on this organic journey of about three decades and thinking of the future, it is equally important to understand this work as an initiative towards building up resilience in village ecosystems with people an integral part of it. This work has taught that the resilience of a village ecosystem, and for that matter community resilience, is not just about resistance, accommodating to or recovering from

A. Pande (✉) · L. Pande
Uttarakhand Seva Nidhi Paryavaran Shiksha Sansthan (USNPSS; l),
Almora, Uttarakhand, India
e-mail: konya.erika@uni-eszterhazy.hu

E. Pénzesné Kónya et al. (eds.), *Environmental Sustainability Education for a Changing World*, https://doi.org/10.1007/978-3-030-66384-1_3

stress/shock to return to a pre-stress position but about moving towards a new and unknown state of balance between people and their environment (Jackson 2005a, b).

Our context is the state of Uttarakhand, which is located in the central Himalayan mountain range in the north of India. Covering an area of about 53 thousand square km, the state has a population of about ten million (Uttarakhand Directorate of Economics and Statistics 2018). It is bounded by Nepal in the east and by Tibet in the north. Being the youngest mountain system of the world and still growing, the Himalayas exhibit innumerable signs of geomorphological fragility and ecological, socio-economic vulnerabilities (USNPSS 2017; Valdiya 2013). The rugged terrain is susceptible to natural hazards, seismic activities and shows extreme and unpredictable weather patterns. Steep micro-climatic differences in temperature, rainfall and soil moisture nurture multiple layers of rich bio-diversity niches in the region but also create conditions for acclimatization of certain species to limited altitudes. More than half of the total population lives in rural areas with agriculture being their main occupation. Traditional practices such as growing different crops at different altitudes, cultivating millets in un-irrigated terraced fields, fruit trees on the colder sides of the mountains, keeping the land fallow in winters, migration with cattle and livestock to the high-altitude pastures during summer and to the lowlands during winter are a few examples of taking advantage of differences in the local climate to generate livelihoods.

Scattered hamlets, with an average of about 40 households, dot the mountain slopes. Until recently, most villages were considered "remote" as roads and communication facilities were limited or non-existent. Though people lived in a seemingly rich ecological setting, economic poverty, socio-political isolation from the lowlands and dependence on subsistence agriculture had created a distinct life in communities. In the early 1990s, economic liberalisation in India stimulated a general growth in the socio-economic sector and transportation and communication facilities began to improve. Educational and economic upliftment has since led to greater mobility and diversification of agriculture-based livelihoods into the non-farm sector. Traditional houses, clothes and food items now give way to cement concrete buildings, machine-made clothing and food from outside the villages/region.

One of the oldest towns in the state of Uttarakhand, Almora, had served as the capital of the local kings before the British regime took over governance in 1815. Later, during the independence movement, Almora witnessed a lot of political activism; several freedom fighters had their homes and relations in this old town. However, despite its historical and political significance, until recently, the town retained its quiet and "sleepy" character; it remained the cultural hub of the state but never gave way to the consumerist tourism-based growth witnessed by neighbouring towns. It is only recently that new flat roofed buildings and hotels have sprung up to stand in line with the traditional slanting slate roofs, stone-mud houses. Today, the people who flock to the market speak their local Kumaoni dialect mixed with Hindi, the national language. Meanwhile, younger generations, not least those many students at the local university campus, increasingly use English and think it fashionable to emulate manners picked up from television channels, western albums and movies. The older generation remains nostalgic about the glory of their past and spends its time discussing the political affairs of the country and beyond in roadside shops and tea-stalls. Of course, this is what Almora is famous for—it is said that everyone in the town is an intellectual and will speak on any issue for hours! In the town centre stands the office of the Uttarakhand Environment Education Centre or USNPSS (Seva Nidhi), as it is called locally. This is located in a traditional stone house; one of the few surviving old buildings that remain functional. The house exudes positivity in all its architectural details and blends elegantly with its surroundings to create an ambience that is typical of Almora. (The restoration of this, formerly dilapidated building and its conversion into office space using local inputs, both human and material, was a labour of love that took 2 years, 1994–95.)

The USNPSS mission is to develop, through education and action, cohesive communities empowered to create rich, sustainable lives for themselves and for future generations. For many years, the USNPSS has served as the nodal agency of the Department of Education and the Ministry of Environment, Forests and Climate Change, Government of India, for the state of Uttarakhand. The environmental education programme and the pre-school programme developed and executed by USNPSS have been mainstreamed in the government system. The organisation also runs one of the oldest and the largest networks of rural women's groups (Uttarakhand Women's Federation) in the hill districts of the state. More details can be found on the USNPSS website www.sevanidhi.org.

2 Working with the USNPSS: A Personal Testimony

Discussion of the work of the USNPSS begins with a first hand experiential account of the way that its lead author, Anuradhe Pande, became involved. Anuradhe writes: "I first learned of the USNPSS in 1989 when my professor sent me to Almora for an assignment in geology. I met another professor in 1991, who, by chance, asked if I knew about an organisation in Uttarakhand that, he said, had started an innovative programme on environment education and community development. Since Almora was situated about 50 kms from my home town, I decided to visit USNPSS on my next trip home.

I joined USNPSS in late 1991. Within a few days, I was taken on a tour to villages. The director himself accompanied me along with another colleague. Together, we travelled to remote areas where some individuals and small community-based organisations (CBOs) had created informal learning centres for 3–6-year-old children. At the time, the USNPSS was in the process of designing and writing textbooks on environmental education for government schools (Pande 2001, 2002). Some government teachers and principals were involved in teaching lessons to grades 9–10 and through them USNPSS received continuous feedback from students and the villages where the course was taught.

Our journey involved walking many miles to reach villages and schools to interact with women, students, teachers and principals. Against the backdrop of the snow-bound peaks of the Himalayas, green pastures, beautiful rivers and high waterfalls, these villages seemed peaceful but below that outwardly placid exterior lay a vibrant and complex society with a multitude of dynamic inter-relationships between the people themselves and between these people and their surrounding habitat and its natural resources.

During these visits, the director would ask me about my impressions and feedback. This was not easy. This young geologist, with her new post graduate diploma on environmental management and year's training in environmental education, had her own ideas/theories on how issues should be addressed. Her thoughts were guided by the books, journals that she had read. Terms like biodiversity, sustainable development, community participation, watershed development, Stockholm Declaration (1973), Brundtland Commission Report (1987), were part of her vocabulary. "What do these terms mean here in the village? What do people make of it when you say we must conserve/preserve bio-diversity? Is there a connection between forests and the farms? Who is cutting the trees—the government, the mafia or the local women (and men)?" In the backdrop of Uttarakhand's 'Chipko' movement of the mid and later 1970s (a Gandhian-led movement for environmental protection that was the origin of the international 'hug the trees' movement but which grew out of an informal attempt by village women to protect their livelihoods and forest resources), people were passionate about issues of conservation and the protection of natural resources. While this process was not bereft of contradictions, everywhere in Uttarakhand, the debate on the movement and its impact were both very lively and yet sensitive (Aryal 1994; Mawdsely 1998).

Responding to these questions was difficult, yet it made me appreciate the fact that some alternative thought and ideas do exist and that bookish knowledge/academic research is not the only way to understand the issues I was seeking to explore. The simplicity of rural life concealed its complex relationship with the surrounding natural resources, the rich mountain biodiversity concealed its current degenerated status, the rich ecological wisdom in the village communities concealed powerful aspirations to be a part of the mainstream development discourse, the apparent economic poverty in these rural communities concealed great generosity. The rural landscape offered so many contradictory yet interesting issues. A colleague advised, "Unlearn what you have learnt in college and go to villages with an open mind and you will see how different things are here". Slowly, the value of this approach became apparent and helped me to appreciate the deep interconnections among people, cattle, water and the land in village communities. This experience also directed me to work with the boundaries of my own limitations regarding rural life.

More visits to villages followed and one of the most vibrant aspects of this exposure was face-to-face interactions with rural women. In each village we went to, we could find a few elderly women who were courageous, very bold and articulate. However, the majority would be very quiet and say little. Finding this diversity of personalities within a small community was puzzling. What was the reason for this variety in behaviour in such similar socio-ecological circumstances? How could it be, within such a community, a particular illiterate woman belonging to a poor household or representing a socially disadvantaged section could be so clear in her mind that the others would come to her for advice/consultation? Cutting across caste, education, class boundaries all women in a community worked very hard as they were responsible for fetching fodder, fuel wood, water and did all agriculture related work, besides performing the household chores such as cleaning, cooking, looking after the children and the elderly in the family. However, some of them conveyed such a strong sense of personal strength and belief in the collective power of women that the experience would reverberate in me for months. From these interactions with women in remote villages, I began to wonder what leadership meant in such a community? How did gender affect the processes of seeking freedom and justice in common people's lives? In these circumstances, what would ecological consciousness mean and how would it facilitate or hinder the processes of social change?

This chapter explores the journey of USNPSS and myself within it, as it has evolved over the last three decades, in seeking meaning in mountain people's lives, both in theory and practice. We describe how our focus on gender, environment and education led to an organic evolution of thoughts and activities that could transform educational policies of the state government and, at the same time, brought in significant changes in community practice and values."

3 Reflections on Practice of Transformation

From the beginning, a holistic view to development had required simultaneous implementation of various programmes bringing together the central and the state government, schools, village residents, motivated individuals and CBOs to collaborate and learn from each other. In the process, answers to some of the questions raised above became apparent whilst others still inspire me to explore, adapt and change with time. The canvas for exploration and expression of these experiences is ever changing both in communities as well as outside; be it the change in theories and practice of environment and development in the national and international arena or changing aspirations in communities.

Four issues bear special relevance for this chapter. First, how could the USNPSS integrate issues of education, environment and development and ensure funding for the programme? Related to this enquiry is the issue of how the organisation could be sustained and advanced and how to create a niche for itself? What worked as the cushion against times when funding was lean or stopped abruptly? Second, what is the value of designing a socially relevant and academically stimulating curriculum on environmental education and of connecting pedagogy so that there was no incongruity between formal and informal learning systems? The third pertains to the ways in which rural communities, especially women, have begun to "think differently" and initiated activities for the collective good. Several questions emerge, such as how to create an enabling environment for people to gain self-confidence and take up activities for their communities, what would leadership mean for women and what challenges would they face when stepping away from their routine traditional duties? The fourth element offers some insights that have emerged from reflection on evolving practice in the USNPSS itself; which is grounded in the concepts of (a) looking at mountain villages as unique ecosystems, (b) transformative education developing from lived-experiences of communities (e.g. Jackson 2008) and (c) feminism developing from the changing relationships of women (and men) with their surrounding natural resources and increasing exposure to the outside world.

We begin with the issue of funding because it compelled us to define the term 'environment' on our own. It required us to explain to the Government of India, why we requested grants for schools and their communities from a single programme. The real challenge was that separate ministries, each with its separate schemes and policies, were responsible for education, environment, rural development and the sectoral ethos would overpower any possibilities of collaboration in bureaucracy. The education department could offer grants for schools or informal learning centres for children but how could one explain to them that the idea of women's empowerment or them taking small projects to improve the well-being of their communities should be under the same programme? Our participatory consultations in villages, especially with women, highlighted the necessity of first meeting their practical basic needs for fuelwood, fodder, access to water, installing sanitation facilities but, in public discourse, any reference to the environment or community development was contradicted with a reminder that education and schooling were separate issues. The challenge was to reconcile the actual needs of people with government funding and discourage any disconnect between the two.

The USNPSS took up an inclusive socio-ecological perspective to highlight connections between various elements that constitute life in the mountains. This perspective eased the argument for supporting practices that, in conventional thinking, would fall outside the boundaries of "education". Eventually, the education department accepted the USNPSS "extended view" of environment and education as connected and interdependent issues. For example, support for the installation of sanitation facilities was sought by advancement of the ideas of (a) "learning by doing" and (b) establishing linkages between the toilets and access to water for cleaning. So, the installation of a toilet had a learning process-based approach, where the people would talk through the work involved in regular monthly meetings and construct the facility on their own (with partial assistance offered by USNPSS). Of course, such activities also interface with a health-education approach. The communities appreciated the educational component and people became knowledgeable about the fact that a lack of sanitation facilities could cause water-borne diseases viz. jaundice, typhoid, diarrhea and dysentery. Soon, women and girls began to appreciate this activity because a facility near/in the house not only offered privacy during periods and pregnancy but also improved a sense of safety as they did not have to go out in the dark where there was fear of a prowling big cat or wild boars.

The programme had ripple effects in villages and soon the news spread all over the mountains. There were no cell phones, television or other equipment to connect with each other but people would exchange ideas. While visiting her parent, a young mother would talk about the informal learning centre that her children were attending or a school teacher might want a centre in his own village and so on. Often, individuals, who 'wanted to do something' for their communities, would contact the USNPSS. Soon, a dynamic network of self-motivated individuals, CBOs, women's groups and youth groups began to function all over the mountain villages of Uttarakhand.

Meanwhile, the USNPSS was designing, testing and developing textbooks for grades 6–8 on environment education for government schools. Five-day residential training programmes were developed and organised in a school for government in-service teachers, since we did not have a place of our own to conduct trainings. In addition, orientation workshops/review meetings were held for the principals and officials in the local administration (for example the school inspectors). This local-specific course on environmental education required students and teachers to select and visit a nearby village to collect data on availability and consumption of natural resources and see if it was sustainable. The next step was to rehabilitate the village ecosystem by improving the availability of forests, water, fodder, fuelwood etc. to ensure sustainability. The programme established synergy between schools and communities as students and teachers had to make several visits to villages to talk with the elderly and to quantify the availability and consumption of natural resources.

The curriculum for children attending pre-school centres had a rich environmental component too. Locally relevant, stimulating activities using clay, local seeds and cereals, leaves, water, waste paper etc. helped shape awareness of their own surroundings and in strengthening eye-hand coordination and motor skills in early stages of children's lives. A curriculum rich in environment education was helped lay down foundations for mental, social, physical development. Environment education could also foster imagination and creativity in a contextualized experience of both personal and collective development of children.

Congruity between formal and informal learning spaces had an additional advantage of conceptualizing education as "a whole" in multiple levels of the self and the collective, as the students, teachers/facilitators and communities all worked together under the same framework. An awareness of the consequences of economic poverty and environmental degradation compelled ordinary people to think and take some action to solve their problems. The real feature of this endeavour lay in reviving and rejuvenating deeply relational cognizance of the self with others and the surrounding environment.

Receptivity to change did not come automatically. It required continuous dialogue facilitated by USNPSS and CBOs with communities, especially adult women and children. It also incurred some expenses regarding travel, food, stipend to the CBOs and field workers. As the programme expanded, space for an office and training centre was acquired in 1994–95 in Almora.

Direct dialogue with hundreds and hundreds of people spread all over the mountains made it evident that, against the general trend of environmental degradation, lay hope and a desire to improve the situation. In villages, much of the daily life was driven by poverty, marginalization, patriarchy and inequalities caused by caste, class and educational background (a sizeable proportion of women were illiterate) and yet people were honest, hard-working and caring towards others. The real paradox was that narratives of human ingenuity and personal struggles were undermined by a bureaucratic and academic discourse, which used its own language and jargon to justify project-outcomes and theories on development. Although, hidden from the official view, these narratives of personal struggles and the brave steps of individuals to overcome barriers soon became a source of inspiration to others involved in the programme and gave enormous strength to the approach USNPSS had adopted. Later, we also learnt that the writings of the late K.E. Boulding (1978) reflect the same perspectives where he emphasises the importance of counter-narratives in developing alternatives to mainstream views of social movements.

In a way, direct collaborative dialogue and constant reflection became the foundation of our own understanding of communities and also of their impressions of the world outside. Open meetings in villages also served as forums to screen slideshows or to explain issues of health, nutrition and well-being to people with the help of posters and charts. In addition, by the mid-1990s, USNPSS had published about 60 booklets, flip charts etc. in simple Hindi language. These booklets were used by teachers, students, the local administration, CBOs and others as they covered a range of themes on subjects of local interest such as the forests, soil, water, trees, agriculture, horticulture, education, health of women and girls, livelihoods etc. (e.g. USNPSS 1993, 2001, 2009).

Appreciation of such "hidden talents, views and struggles" led communities to believe that their thoughts and experiences were valuable and that they needed to talk about issues, which in conventional thinking would not allow them to do so. This approach also helped in bridging the gap between the private and the public domain. For example, a process of representing the unrepresentable commenced as women got the courage to talk about issues of abuse and violence against them by men under the influence of alcohol, and took action. For the first time in communities the unrecorded was recorded, which helped women talk about "private family matters" and so make sense of what was happening with their own selves, their families and beyond (Fig. 3.1).

4 Transformation of the Self

Personal and collective responses to the stress arising from marginalisation, economic and ecological poverty did change community practices and led to a higher level of environmental consciousness. Transformation in the self was reflected on various fronts such as changing attitudes and behaviour in daily life. Some individuals came forward to work in villages, sometimes leaving behind good careers. In communities, individuals began to shed personal inhibitions and developed the confidence to speak-up. Women began to change their attitudes and behaviour regarding the traditional practices that were observed during menstruation and pregnancy. Taking a bath and washing clothes every day, especially during periods, was one such change that reduced incidences of leucorrhoea among women. Slowly, the old

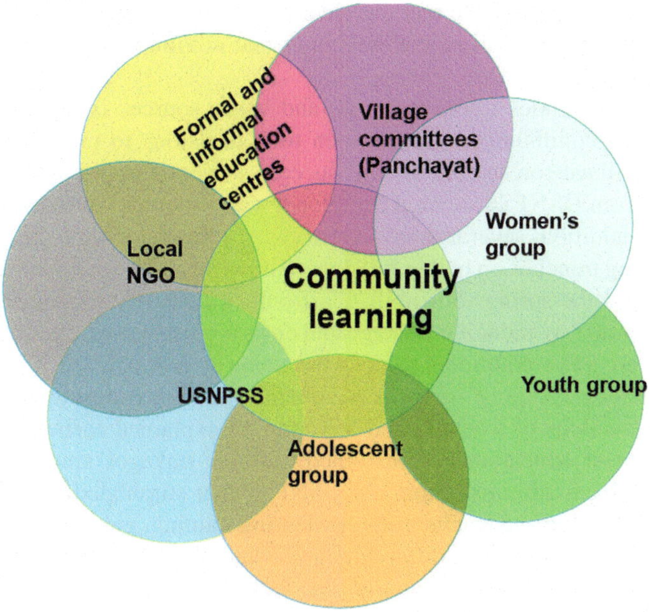

Fig. 3.1 Community learning requires the participation of all village residents as well as formal and informal institutions in the community. It is not about creating a champion who solves problems for others but to nurture a movement where everyone participates.

practice of girls and women being secluded for 5 days during menstruation discontinued. In addition, clothes, bedding, the rooms, the village paths all became cleaner at once.

The transformative process entailed in community learning has been empowering. The adolescents, youth and adults began to understand the socio-political and economic systems and their implications in personal and collective life. For example, women learnt how the structures of patriarchy and caste would affect their personal position in the community and how they could create space for themselves, change and move forward. An adolescent girl would learn to understand her rights and exercise them in daily life.

5 Collective Transformation

Informal learning centres moved beyond the boundaries of just providing access to education for children towards integrating everyone in the community to learn and change. For example, conventional custom would dissuade most girls from attending schools. However, the informal learning centres encouraged girl's education (Pande 2015). All girls in a village were enrolled in the centres and continuous dialogue with their parents and others ensured regular attendance. Further, the facilitators would make sure that the older girls attended school on a regular basis. Today, elementary education is almost universal in villages and a sizeable proportion of girls attend college.

Change in attitudes towards forests and water sources is another example of change at the community level. Women took decisions to protect, conserve and regenerate village forests and several success stories emerged. With some partial assistance from USNPSS, several traditional water sources were restored and rejuvenated. In addition, several innovative ways of rain water harvesting were developed. Digging trenches on higher hillslopes allowed water to percolate in the ground and recharged the springs downhill. Plastic-lined tanks for roof-water harvesting, conserving water by reducing wastage and ensuring equal distribution to all in communities were a few additional steps in this same direction.

The pedagogical transformation in schools synergized development of several innovative methods of teaching and learning. As explained earlier, the curriculum on environment education had instituted multiple linkages among communities, schools, the local administration and USNPSS that stimulated innovation in both methods of learning and teacher training. During training, government school teachers were challenged to think differently and come out with ideas that would benefit both their students and their communities. A difficult task as it was, teachers would request USNPSS to "tell us what to do and we will do it." This proposition was however, rejected by USNPSS who, instead urged these teachers to think for themselves. To encourage this practice of self-thought and self-direction among teachers and the principals, innovative ideas and practices were integrated in the curriculum and in training and orientation programmes. This practice not only gave a boost to their confidence but also motivated the principals and teachers to think deeper about issues of environment and socio-ecological situations at the local level.

As a reform in policy, the state government mainstreamed the environmental education programme in all government schools in the year 2002 in grades 6–8 (www.sevanidhi.org). The course was studied by about half a million students who attended government schools. The government also took over the task of printing and distribution of the textbooks developed by USNPSS. At another level, the curriculum developed for the informal learning centres was also taken up by the state under its 'Early Childhood Care and Education' programme (Government of Uttarakhand 2003) and, eventually, mainstreamed in government run centres.

A major setback to the programme occurred in 2005–06 when the government scheme under which USNPSS was receiving grants discontinued abruptly. Besides USNPSS, the jolt was felt in far flung villages where informal learning centres and women's groups were functioning. Subsequently, USNPSS started to look for funding elsewhere and eventually received grants from some charitable trusts and donations from individual well-wishers. This financial shock, however, forced USNPSS to reduce its activities and the physical coverage of the programme.

What do all these changes in multiple levels mean in view of theories and practice of transformative learning and in building resilience in communities? Four issues seem key: collaborative discourse and its challenges, practicalities of strengthening transformative education, avoiding self-admiration and self-serving purpose, and ecosystem resilience as basic elements to develop communities of practice.

6 Collaborative Discourse and Its Challenges

Creating an enabling environment and space for dialogue was the first step in the process that led to an organic evolution of practices of transformative learning. This process gave voice to, and built confidence in, people who had very little say in practice. The long-term association of USNPSS with communities not only developed relationships with people, their resources (forests, water sources, cultivated fields etc.) and the institutions (schools, village committees etc.) but also nurtured respect for local customs and culture. However, certain customs, values and traditions had either lost relevance or become a hindrance in current times, therefore change was required. For example, practices that positioned women and other socially disadvantaged sections of the population at the margins needed to be questioned and worked upon. The challenge was to find a way of approaching such issues that would enable critical reflection in villages and, perhaps, lead to positive change.

Meeting this challenge was tough. In still pristine villages, people led a simple life and sustained their needs from surrounding resources. However, the literature on environment, sustainability, ecological services (Costanza et al. 1997), and participatory development confirmed our ideas of strengthening the village ecosystems, especially their natural resources/biodiversity, to develop resilience in communities (Gunderson 2000; Drever et al. 2006; Ruhl and Salzman 2006; Gunderson et al. 2010). The Millennium Ecosystem Assessment 2005 had referred to resilience as *the amount of stress or disturbance that an ecosystem can absorb and still remain capable of returning to its pre-disturbance state.* Combining insights from the literature with our own understanding of the grassroots realities, we at USNPSS believed that building resilience would mean strengthening the natural resource base as well as community co-operation and solidarity so that the village ecosystem could absorb or adjust to any amount of pressure that it was subjected to and recover to its original state. Briefly, a village ecosystem in Uttarakhand is governed by the interdependence that exists between the land (cultivated and not cultivated), water, cattle and people along with the flow of energy—sun as well as human labour (Jackson, 2005a; USNPSS 2014, 2017). Improving the natural resource base to improve the health of the ecosystem is therefore an essential step towards building resilience in village ecosystems.

From the late 1990s, networks of roads started developing in the state. Next, television and cell phone facilities developed so improving both mobility and exposure to the outside world. By this time, USNPSS was working in about 450 villages and almost all girls (and boys) were enrolled in schools. The young educated children could use the cell phones easily; songs/video downloads in cell-phones became the source of information/entertainment in villages.

Soon, criticism of globalisation, structural adjustment policies and of western dominated ideas of cultural homogenisation began to appear in literature (e.g. Follen and Clover 1997). This new wave was apprehensive of the impacts of the global applications of ideas and rapid changes on the culture of communities; but what we observed in these villages was a high receptivity to such changes. Rural people in Uttarakhand, totally missed out the phase of having landline or telephone booths for public service and direct access to and ownership of cell phones soon became first a status symbol and later a necessity. Schooling and such exposure brought in greater use of the Hindi language and each girl or boy began to aspire for proficiency in English.

Amidst this exposure to the outside world and increasing consumerism lay the community structure with its old customs, traditional values and practices, often creating distress and confusion between what was perceived as the "modern and desirable" and the "traditional and backward". Soon, the staff of USNPSS realised that messages coming from media and from the cities were far more powerful and had more influence on the young men and women than they had ever imagined. Parallel to the stream of education and development ethos that we had designed and worked on for years, another current of mainstream "development" had emerged. This new wave of "development" was symbolized by construction, ownership of consumer goods and it promised a better life to everyone. The older strands of environmental education that advocated conservation and sustainability of natural resources were soon bypassed by people making (or aspiring for) quick money by illegal/legal sand-mining of the river beds, selling liquor or by taking loans under different schemes instituted by the government and the banks.

This new wave of consumerism hurriedly began to excavate sand from the narrow gorges of the mountain rivers, pumped water from traditional village-owned sources to feed the tourists in resorts/hotels with no or little attempt to conserve, replaced the local architecture by concrete-based construction and superceded local ways of dressing and of food habits even in remote areas. As the mindset towards life and standards of living changed, increasingly the young people began to flee to the cities to find some job/work that would generate cash (Mehta 2014). Those who could not go out began to move to the roadsides and opened grocery or tea shops.

The whole debate, therefore, now shifted to an increasingly contested emphasis on whether to succumb to the capitalist way of development or to continue with the ideas of environmental sustainability, participatory development, village self-sufficiency, and so on. This discourse required much thinking, planning and a vision for the future. In its eagerness to find some solution to these conflicting issues, USNPSS started to conduct research and publication to bring into the public domain the changes and the confusion communities were experiencing (USNPSS 2014, 2017). The organisation also started to learn more from scholarly work on environment and development at the international level, both by meeting experts who would visit and through the media. The process was enriched by discussing such ideas with CBOs and women's groups who offered practical advice and real-life experience to this debate.

7 Practicalities of Transformative Education

In a way, the changes occurring in villages and the quest to find solutions to those issues brought us to a critical level of thinking about how to strike a balance between the "traditional" and the "modern". With women's groups and CBOs, we have now begun to discuss development in relation to marketing decisions made outside in which they have very little choice or control or to climate change issues that affect them the most in lieu of their location in a fragile mountain system. The same meetings also take up issues of conditioning and control by gender, caste, class etc. and how it influences relationships at home, in the community and at regional-national levels. In addition, the inter-subjectivity of our thinking has brought into focus the differences as well as sameness in lived experiences so supporting change in a positive direction.

A direct outcome of this engagement with the world outside and collaborative dialogue at local levels is innovation and integration of new ideas/practices at multiple levels. Experience early in the programme had established that an energetic and cognitive way of change in one person and their community may lead to an organic growth of ideas in others too. Taking our cue from this insight, we began to think of a convergence model in 2014. Till then, programmes with children, women, adolescent, youth and in schools were all self-directing and self-fulfilling with operational linkages amongst them. While maintaining the self-directing goal, we began to converge all these programmes into one and called it community learning through transformative education. About 100 libraries and learning centres that were functional during that time in villages were brought under one umbrella by unifying them with projects taken up by women's groups, adolescents and youth groups. The centres were then called Village Learning Centres (VLCs). In practical terms, we began to hold monthly meetings of women's groups and adolescent girls' groups in the centre itself. Children, 6–14 years old, attended the centre after-school and a new curriculum was developed. Learning activities focused on improving knowledge and skills of mathematics, language, environmental education, general awareness regarding the self, the community, the region and national/international trends in development.

It was the relational aspect of consensus building for action in these communities that led to change in thoughts and practice. For the USNPSS, building up relationships has meant involving all in the community who offer support, work together for collective good and change in society.

In this process, the USNPSS had constructed the idea of working *with* communities, not working *for* communities, and actively developed the ability of listening to people's voices. This requires us to connect with communities in ways that instill confidence through dialogue and action. We believe that everyone is unique yet vulnerable and imperfect in many ways in her/his humanness. We do not go to the villages with an idea of "fixing their problems" or "fixing their selves" as if people are deficient and need fixing by us. As we listen more and more to real-life experiences at multiple levels, the work becomes more responsive and accountable towards communities.

8 Avoiding Self-Admiration and Self-Serving Purpose

Creating a genuine ethos of avoiding self-admiration and self-serving motives had meant that USNPSS could focus on action without worrying about publicity for its individual worker or the organisation. Indeed, the organisation has never made any conscious attempt to seek publicity and, many times, refused to engage with people who would come for a day or two with an offer to publicise but without going deeply into the thoughts and practice of this work. In a way, this approach helped the USNPSS concentrate on its work while not wasting time and energy in pleasing others. Yet, support came on its own volition. One of our friends told us that we must have a website and offered to develop it and so he did. Another friend volunteered to make a video film and upload it in our website and so on.

As described, transformative learning builds the basic attributes of humanity which include passion, care, ingenuity, honesty, trust and confidence in oneself and others, but this is not the full picture. The socio-economic structure and the administrative system also regulate one's position in a community. Therefore, the barriers of inequality and of bureaucracy require change or adjustment to accommodate growing aspirations of people. We have dealt with this issue at multiple levels. For example, hundreds of women associated with USNPSS have become elected women representatives in local governance systems. They take up issues of equality and inclusion in village committees and beyond. In addition, women's groups show an alternative approach of organising and working in their communities. Contrary to the mainstream approach of organising women as small groups within a community (for example the self-help groups or the micro-credit projects), all women are seen as members of a single group, a Whole Village Group (WVG). Irrespective of their caste, age, economic and educational status, all the women in the WVG sit together (usually in a circle) and they talk.

This approach has helped not only weaken socio-cultural disparities but also to build a sense of community and co-operation fostering intra-village (as well as inter-village) synergy. It has also helped in dealing with common social problems such as alcoholism and gambling among men. As described, women have gained sufficient confidence to shed social taboos and bring difficult private matters into the public domain. Second, they have been able to make decisions to punish those who were caught abusing others or beating their wives and children under the influence of alcohol. Third, these WVGs have taken collective action against people who defy or not abide by rules that they had made in village meetings. Usually, such a person would have to pay a fine. The funds so collected would be deposited in women's collective fund and this resource of seed money grows over time. Using this money, women's groups have bought large utensils and furniture that they rent out during festivities in their own village and in the neighbourhood. The collective approach has helped these WVGs avoid self-serving motives and strengthened the value of collective efforts and co-operation.

9 Community Resilience

One of the key insights generated from this USNPSS work has been that community resilience must be seen in the context of the resilience of the whole village ecosystem (Berkes et al. 2003; Folke 2016). Resilience and ecosystem health are inextricably linked. By co-creating knowledge with communities, USNPSS could develop new pathways to support and strengthen resilience in both communities and the ecosystems of which they were a part. This involved understanding and defining the boundaries of a village ecosystem as well as the economic, socio-political forces that people experienced in their daily life.

However, recent changes have forced us to think deeper into the issue of resilience. With improving infrastructure, both the inflow and outflow of materials and human resources have increased and the boundaries of a village ecosystem are expanding. As more and more of the outside world enters into the "inner ecosystem" of communities, aspirations of the youth change. Ideas and images of the outside as brought about by increasing mobility, use of television and cell phones and the availability of more options for exercising life choices have now diluted the boundaries of a "closed or self-sufficient model" of a village ecosystem. As people migrate from the villages, the close network of traditional social ties has become weakened. The learning from this change in communities is that resilience is not static but a vibrant and dynamic quality that has a subjective aspect and can be interpreted in different ways in different contexts and situations.

10 Conclusion

The examples of change described in the article point towards the importance of creating space and an enabling environment for rural communities, especially women and children, to respond to the challenges and tensions of development and to find solutions. The article shows how communities can transform from being marginalized, isolated entities into active, vocal groups capable of making decisions to transform their own-self and their society.

For the USNPSS, transformation through education and action formed the core idea around which multiple activities were built to develop theories and practice of community learning. The programme involved women, adolescents, youth, teachers, CBOs and everyone else in these communities so that there could be no disconnect between formal and informal learning systems. The inclusive approach enabled innovations in curriculum and enriched learning practices in government schools as well as in informal learning centres. The process offered new opportunities and opened up new spaces for collaborative dialogue, critique and reflection to people who had never been involved in such endeavours before.

The article shows that building up resilience in villages involves long-term engagement with people and their environment (village ecosystem e.g. Jackson, 2005a, b). Resilience is a distinctive quality of an ecosystem and changes over time. In view of changes occurring at all times in ecosystems, it is difficult for a community to go back to its original position or to resist mainstream changes for long. As described, the intergenerational changes in aspirations of the youth, necessarily force traditional communities to evolve continuously and respond to new challenges. Thus, the approaches for resilience-building do not aim to enable ecosystems to absorb and to recover from stress and so restore their pre-stress positions but to move towards a new state of balance between people and their surroundings. What this new state will be? This is the challenge that USNPSS faces today. It is the 'unknown' of this challenge that inspires us to think of the future, to undertake more experimentation with communities; and to learn from outside scholarship in the field of transformative education, climate change, ecosystems and resilience.

Acknowledgement We are grateful to Professor Martin Haigh for asking us to write and subsequently offering feedback on an early draft of this chapter.

References

Aryal, M. (1994). Axing chipko. *Himal*, 7(1), 8–23.

Berkes, F., Colding, J. and Folke, C. (2003). *Navigating Social-ecological Systems: Building Resilience for Complexity and Change*. Cambridge University Press, Cambridge, UK.

Boulding, K.E. (1978). *Ecodynamics: A New Theory of Societal Evolution*. SAGE Publications, Thosand Oaks, CA.

Brundtland Commission (1987). *Our Common Future, Concept of Sustainable Development*. Oxford University Press, Oxford, UK.

Costanza, R., d'Arge, R., De Groot, R., Farber, S., Grasso, M., Hannon, B., Limburg, K., Naeem, S., O'neill, R.V., Paruelo, J. and Raskin, R.G. (1997). The value of the world's ecosystem services and natural capital. *Nature*, 387(6630), 253–260, https://doi.org/10.1038/387253a0.

Drever, C.R., Peterson, G., Messier, C., Bergeron, Y. and Flannigan, M. (2006). Can forest management based on natural disturbances maintain ecological resilience? *Canadian Journal of Forest Research*, 36(9), 2285–2299, https://doi.org/10.1139/x06-132.

Folke, C. (2016). Resilience (Republished). *Ecology and Society*, 21(4): 44, https://doi.org/10.5751/ES-09088-210444.

Follen, S. and Clover, D.E. (1997). Community revitalization through critical environmental adult education. *In:* SCUTREA (ed). Crossing borders, breaking boundaries: Research in the education of adults. 27th Annual Conference Proceedings 1997. *Agenda* (University of Leeds) 21, 36–39.

Government of Uttarakhand (2003). *Rimjhim – Early Childhood Care and Education, Education for All*, Government of Uttarakhand, Dehradun, Uttarakhand.

Gunderson, L. (2000). Ecological resilience: in theory and application. *Annual Review of Ecological System*, 31, 425–439, https://doi.org/10.1146/annurev.ecolsys.31.1.425.

Gunderson, L.H., Allen, C.R. and Hollings, C.S. (Eds) (2010). *Foundations of Ecological Resilience*. Washington, D.C.: Island Press.

Jackson, M.G. (2005a). *A Future for Rural India: Ecology, Equity, Empowerment*. The Other India Press, Mapusa, Goa.

Jackson, M.G. (2005b). *The Village Ecosystem*. The Other India Press, Mapusa, Goa.

Jackson, M.G. (2008). *Transformative Learning for a New Worldview – Learning to Think Differently*. Palgrave Macmillan, New York.

Mawdsely, E. (1998). After Chipko: From environment to region in Uttaranchal. *Journal of Peasant Studies*, 25(4), 36–54, https://doi.org/10.1080/03066159808438683.

Mehta, M. (2014). *Neither of the Hills nor of the City: Glimpses of Peri-urban Lives in a Time of Rapid Change*. Doon Library and Research Centre, Dehradun, India.

Millennium Ecosystem Assessment (2005). *Ecosystems and Human Well-being: Synthesis*. Island Press, Washington, DC.

Pande, A. (2001). An environmental education course in rural Himalayan schools. *Journal of Environmental Education*, 32(3), 47–53, https://doi.org/10.1080/00958960109599146.

Pande, L. (2002). Our Land Our Life: An innovative approach to environmental education in the Central Himalayas. *In:* Tilbury, D., Stevenson, R.B., Flen, J. and Schreuder, D. (eds), *Education and Sustainability: Responding to the Global Challenge*. Gland, Switzerland and Cambridge: International Union for the Conservation of Nature (IUCN), pp. 65–73.

Pande, A. (2015). *Village Learning Centres in Uttarakhand*. Uttarakhand Seva Nidhi Paryavaran Shiksha Sansthan (USNPSS), Almora, India.

Ruhl, J. B. and Salzman, J. (2006). The Law and Policy Beginnings of Ecosystem Services. *Journal of Land Use and Environmental Law*, 22, 157–172, https://scholarship.law.duke.edu/faculty_scholarship/2409.

Stockholm Declaration (1973). Report of The United Nations Conference on the Human Environment, Stockholm, Sweden 5-16 June 1972. United Nations, New York, https://www.un.org/ga/search/view_doc.asp?symbol=A/CONF.48/14 REV.1.

USNPSS (1993, 2001, 2009). *Hamari Dharti, Hamara Jeevan (Our Land: Our Life Uttarakhand): A practical course in environmental education and agriculture for the schools of Uttarakhand*. Uttarakhand Seva Nidhi Paryavaran Shiksha Sansthan (USNPSS), Almora, India.

USNPSS (2014). Community driven climate resilient hill farming in village ecosystems of NW Himalayas (Uttarakhand). Study supported under the National Initiative on Climate Resilient Agriculture (NICRA) by Central Research Institute for Dryland Agriculture. ICAR, Government of India, 2011-2014. ICAR, New Delhi. www.sevanidhi.org.

USNPSS (2017). Reframing discourse of gender and disaster: Pahari narratives of change from rural Uttarakhand. Study supported under the project titled: Gender Issues in the Development of Ecologically Fragile Zones: A Case Study of Village Communities in Uttarakhand by the ICSSR. Government of India, 2014-2017. ICSSR, New Delhi. www.sevanidhi.org.

Uttarakhand Directorate of Economics and Statistics (2018). *Uttarakhand at a Glance 2016-2018*. Government of Uttarakhand, Dehradun. http://des.uk.gov.in/pages/display/61-uttarakhand-at-a-glance.

Valdiya, K.S. (2013). *Environmental Geology: Ecology, Resource and Hazard Management*, 2nd edition. McGraw Hill Education (India), New Delhi.

Chapter 4
Influence of a Disaster on the Environmental Attitudes of University Students in Uttarakhand, India

Martin Haigh

1 Introduction

Effective headwater protection and management requires community support and engagement. This requires an understanding of the attitudes and values of the host community (e.g. Pande and Pande 2020). This chapter examines environmental attitudes among undergraduate learners at a Hindi-medium university in the headwaters of the Uttarakhand Himalaya, India, in the aftermath of a major disaster. It uses the standard 'New Ecological Paradigm' questionnaire (Dunlap 2008; Dunlap et al. 2000) in its 15-item updated version (cf. Dunlap and Van Liere 1978). It also uses some additional items to explore how local university students in this headwater community think and feel about their environmental circumstances in the immediate aftermath of an environmental disaster involving catastrophic rainfall, flooding and landslides. It considers how these responses may be interpreted and how they are affected by the learners' disciplinary course of studies. Then, more controversially, these response patterns are compared with NEP data recorded elsewhere beginning with data collected at the author's home university and then NEP results collected in other parts of the world. Finally, it considers how differences in cultural attitudes to the environment affect the validity of NEP-style international questionnaires from Western traditions and how a more culturally sensitive questionnaire survey might be developed and employed to enhance environmental education in this Himalayan State.

M. Haigh (✉)
Department of Social Sciences (Geography), Oxford Brookes University, Oxford, UK
e-mail: mhaigh@brookes.ac.uk

E. Pénzesné Kónya et al. (eds.), *Environmental Sustainability Education for a Changing World*, https://doi.org/10.1007/978-3-030-66384-1_4

2 Context

During September 18–19th, 2010, the central part of Uttarakhand State in India's Lesser Himalaya received rainfall, unprecedented in 60 years of record. A late Monsoon storm deposited 277 mm of rain at intensities that often exceeded 33 mm/ h^{-1}. This down-pour followed 2 days of heavy rain, which had already saturated soil and raised water-table levels. Inevitably, the consequence was landsliding and flooding; the River Kosi developed an unprecedented flood surge with discharges reaching 618.1 m^3/s. Meanwhile, the road network, where it was not washed away, was blocked or undermined by landslides. Much remained impassable for weeks. This disaster became headline news across India.

By coincidence, in October, 2010, the author arrived on a mission to explore the environmental attitudes, beliefs and values of Indian university students. However, this mission was, immediately, diverted into work on the diagnosis of local landslide problems. This study was built on previous research that had evaluated landslide causation on the new highway that ran along the contour of a steep, north-facing, lesser Himalayan mountain ridge and through the host college's campus (Haigh and Rawat 2011, 2012; Haigh et al. 1988).

On completion of the landslide field-survey, colleagues at the host university, kindly lent their support to the original educational research project. Two groups of learners were mustered to complete the project's questionnaires and both were supported by help with translation between English and the Hindi medium used in almost all teaching. The two groups were 50 first-year-postgraduate geographers studying a nationally-supported Geographical Information Systems curriculum. The second were an upper level class of 37, mainly local, upper-level undergraduate learners who were training to be teachers. The responses of this second group may have been especially affected by the disaster of the previous month, which had damaged the Education faculty building in which the survey took place. This had suffered cracking of the walls, the collapse of part of the main entry way, and the appearance of a landslide scar immediately downslope. Additionally, during the questionnaire survey, the building was undergoing damage and safety inspection by a team of experts led by the Vice Chancellor (Fig. 4.1).

3 Methods: The Source Questionnaires

Previous research has suggested that those most directly affected by an environmental problem are also those most concerned about that problem, even if factors such as income and education affect their awareness of solution options (Anderson et al. 2007). So, this study examined the null-hypothesis that the environmental attitudes of learners in geography and education are universal and unaffected by location, culture and recent events, no matter how traumatic. Over the years, many questionnaires, perhaps hundreds, have been developed for the measurement of

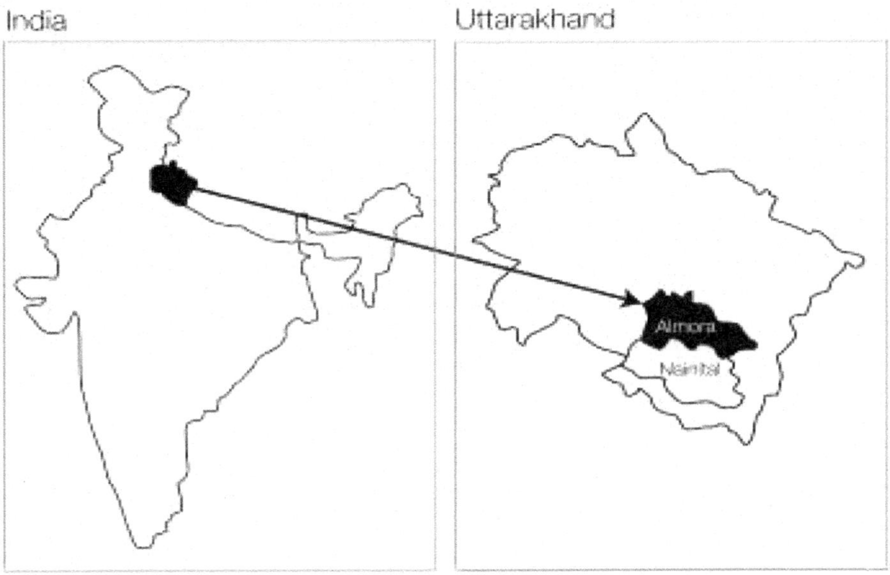

Fig. 4.1 Location map of Uttarakhand

environmental attitudes and this lack of standardisation may contribute to the slow advancement of research into environmental attitudes (Dunlap and Jones 2002; Bird 2009). Among the best known and most widely used of these questionnaires is the New Ecological Paradigm (NEP) questionnaire, which attempts to measure general beliefs about the ways humans think about the environment (Dunlap et al. 2000; McIntyre and Milfont 2016). However, the NEP is not without problems, many of which arise from its roots in the thinking and culture of the USA in the 1970s. It was created as a reaction to the outdated ‘dominant social paradigm’ of those times, which regarded the environment either as a treasure box of resources or a wildness that needed to be tamed (Pirages and Ehrlich 1974). Its original aim was to chart the degree to which different parts of American society had moved away from these attitudes and toward a new environmental worldview that recognised that there were limits to the number of humans that the world could support, that human societies had the capacity to upset the balance of nature, and that contested both the right and capability of humans to govern the natural environment (Dunlap and Van Liere 1978). After much critical exploration and research, a revised and slightly expanded version of the NEP questionnaire appeared in 2000 (Dunlap et al. 2000), which has also been used in cross-cultural contexts (e.g. Erdoğan 2009; Ogunbode 2013). However, the final questionnaire remains more ‘what’ than ‘why’; it does not much address issues of motivation, ethics, spirituality or, most crucially, how such attitudes convert into subsequent behaviour (Schultz and Zelezny 1999; Lundmark 2007; Sita Ramadas et al. 2014). Indeed, recent psychometric research suggests that perceptions of environmental impacts may be unrelated to behaviour intentions, which are mitigated by perceptions of cost and societal pressure as much as by human and environmental welfare (Truelove and Gillis 2018). Designing such questionnaires to elicit and explore multiple perspectives and/or different world views is a relatively recent concern (e.g. Scolobig et al. 2016).

The 15 items of the NEP are displayed in Table 4.1. The questionnaire aims to address five aspects of a pro-environmental world view and three questions address each aspect. These five 'aspects' are: the limits to population growth ('Limits' questions: NEP1, 6, 11), 'Anti-anthropocentrism' (NEP2, 7, 12), ecological fragility and the human impact on the balance of Nature ('Balance' NEP3, 8, 13), the risk of an ecological crisis ('Ecocrisis' NEP5, 10, 15) and the notion that human ingenuity can solve all problems ('Anti-exemptionalism' NEP4, 9, 14) (Dunlap et al. 2000). Modern concerns about climate change are not directly addressed, although, of course, they are relevant in this case (Helm et al. 2018).

Each item is measured on a Likert scale ranging from 5, meaning strongly agree, through 4 meaning mildly agree, 3 meaning unsure or don't know, 2 meaning mildly disagree to 1 meaning strongly disagree. In the questionnaire, agreement with the eight odd-numbered questions is supposed to indicate pro-environmental attitudes while agreement with the seven even-numbered items indicates greater anthropocentrism. Normally, the scores for these seven items are reversed for statistical analysis; a score of 3 then becomes the boundary between an anthropocentric and pro-ecological worldview. Pro-environmental attitudes are expected to elicit high scores while the more anthropocentric responses of the so-called 'dominant social paradigm' are expected to be low. Overall, around 30% of all NEP studies concern college students (Hawcroft and Milfont 2010). Recorded means lean towards the positive but overall they range from 2.95 (Nigeria) to 4.11 (Canada) (Ogunbode 2013; Schultz and Zelezny 1999).

Here, scores from the seven even-numbered items are left unchanged in several tabulations. This is because, all that is really certain is that, the respondents answered the questions as asked and the scale(s) imposed by the designers may not be entirely apt. These tables are described as containing 'raw' scores.

Here, the revised NEP questionnaires, together with the supplementary questions, were administered during class-time. Each question was provided both in English and in Hindi translation, for those less comfortable with the English language. Students responded for themselves but, typically, after discussion about the question's meaning with their classmates and immediate neighbours. The questionnaires were administered to the two groups of learners on consecutive days. During the survey, staff-student interactions were much greater amongst the geography contingent and discussions about each question's meaning were much more widespread. By contrast, the Education survey was conducted in a quieter, more 'schoolroom' style, with less staff interaction and questioning but, perhaps, more discussion between neighbouring classmates. Class time given to each survey varied between 35 and 45 min.

Table 4.1 The NEP (New Ecological Paradigm) Questionnaire and Raw (i.e. even numbered questions not reverse recoded) Scores. Comparison of Geography vs. Education and male vs. female Uttarakhand students

Uttarakhand University	NEP mean	Std. dev.	Geography mean	Std. dev.	Education mean	Std. dev.	Mann Whitney-U Sig	Male mean	Std. dev.	Female mean	Std. dev.	Sig Kruskal-Wallace IS	NEP "Aspect"
NEP1. We are approaching the limit of the number of people the Earth can support.	4.00	1.08	3.82	1.23	**4.24**	**.76**	**0.016**	3.98	1.27	4.03	.73		Limits
NEP 2. Humans have the right to modify the natural environment to suit their needs.	2.69	1.25	2.60	1.17	2.81	1.34		2.76	1.31	2.50	1.13		Anti-anthropocentrism
NEP3. When humans interfere with nature it often produces disastrous consequences.	4.40	.99	4.30	.93	**4.54**	**1.06**	**0.003**	4.27	1.04	**4.69**	**.81**	**<0.0005**	Balance
NEP4. Human ingenuity will insure that we do NOT make the earth unlivable.	3.63	1.11	3.47	1.13	**3.83**	**1.05**	**0.046**	3.67	1.13	3.48	1.11		Anti-exemptionalism
NEP5. Humans are severely abusing the environment.	4.28	.71	4.34	.74	4.19	.66		4.27	.72	4.25	.71		Ecocrisis
NEP6. Earth has plenty of natural resources if we just learn how to develop them.	4.40	.80	4.40	.64	4.40	1.00		4.39	.80	4.57	.56		Limits

(continued)

Table 4.1 (continued)

Uttarakhand University	NEP mean	Std. dev.	Geography mean	Std. dev.	Education mean	Std. dev.	Mann Whitney-U Sig	Male mean	Std. dev.	Female mean	Std. dev.	Sig Kruskal-Wallace IS	NEP "Aspect"
NEP7. Plants & animals have as much right as humans to exist.	4.44	.87	**4.64**	**.75**	4.17	.96	**<0.0005**	4.39	.89	4.52	.88		Anti-anthropocentrism
NEP8. Nature is strong enough to cope with the impacts of modern industrial nations.	3.09	1.34	2.76	1.25	**3.54**	1.34	**<0.0005**	3.02	1.40	3.16	1.31		Balance
NEP9. Despite our special abilities humans are still subject to the laws of nature.	4.03	1.11	**4.38**	**.87**	3.57	1.22	**<0.0005**	4.12	1.07	3.88	1.23		Anti-exemptionalism
NEP10. The so-called "ecological crisis" facing humankind has been greatly exaggerated.	3.55	1.20	3.55	1.20	3.54	1.21		3.32	1.36	**3.93**	**.82**	**0.006**	Ecocrisis
NEP11. The Earth is like a spaceship with very limited room and resources.	3.48	1.24	3.16	1.32	**3.92**	**.98**	**<0.0005**	3.51	1.34	3.31	1.08		Limits
NEP12. Humans were meant to rule over the rest of nature.	2.43	1.21	2.20	1.06	**2.73**	**1.33**	**0.012**	2.24	1.17	2.56	1.18		Anti-anthropocentrism

NEP13. The balance of nature is very delicate and easily upset.	4.01	.91	3.98	.74	**4.05**	**1.10**	**0.035**	3.92	1.012	4.16	.76		Balance
NEP14. Humans will eventually learn enough about nature works to be able to control it.	3.92	.92	3.94	.65	3.89	1.19		3.78	1.001	**4.13**	**.79**	**0.022**	Anti-exemptionalism
NEP15. If things continue on their present course, we will soon experience a major ecological catastrophe.	4.30	1.16	4.40	.99	4.17	1.35		4.39	.822	4.26	1.40		Ecocrisis

Significantly more extreme responses are emboldened

4 Results

Table 4.1 explores the small data set ($n = 87$) retrieved from the administration of the New Ecological Paradigm questionnaire to the Uttarakhand college learners. It shows means and standard deviations for the whole sample, for subsamples divided by discipline and gender, and it indicates where there are significant differences in the response patterns of these subsamples. In these tests, data from the even numbered questions has not been recoded and it is assumed that there are no significant differences in the response patterns.

In general, female respondents are thought to record higher scores on the NEP scale than males (Dunlap et al. 2000). However, here, in Table 4.1, there are only three significant differences. The strongest of these is for NEP3, where female learners agree more strongly that human interference in the environment can produce disastrous consequences ($p < 0.0005$). However, in the (unrecoded) NEP10 and NEP14, they also agree that the risk of human actions creating an ecological crisis are exaggerated ($p = 0.006$) and, just possibly that, humans will find a way to control nature to prevent any such thing ($p = 0.022$). Since the NEP questionnaire contains 15 items, this increases the likelihood that a 'significant difference' could be produced by chance alone. Applying a strict Bonferroni Correction would reduce the threshold of significance to just $p = 0.003$ leaving both these differences in the grey area of possible significance between $p = 0.05$ and $p = 0.003$.

Table 4.1 also compares across the two disciplines represented, although it should be remembered that the Education students were also those working in the building that had been damaged by earth movement during the storm. Using the $p = 0.003$ significance threshold, there are two areas where the Geography learners hold stronger beliefs than the Education learners. These both suggest that the Geographers believe more strongly that humans are not separate from Nature; they agree more strongly with NEP7 ($p < 0.0005$) in the right to exist of other creatures and with NEP9 ($p < 0.0005$) that humans remain subject to the laws of Nature. Unsurprisingly, given their immediate environment, the Education learners believed more strongly with NEP3 ($p = 0.003$), that human interference in the environment could produce disastrous consequences. They also gave more support to the notion that the Earth has limited resources (NEP11; $p < 0.0005$) but also that Nature is strong enough to withstand modern industrial society (NEP8; $p < 0.0005$). This was the only situation where the Education learners disagreed with the Geographers, who did not support this statement. In the other areas of possible difference, Education learners expressed more strongly pro-environmental views than the Geographers (NEP1, NEP4, NEP12, and NEP14).

One peer reviewer of this paper suggests that it would have been better to simply present a paper exploring 'the influence of disaster on Environmental Attitude amongst university students in Uttarakhand State' and not attempt to compare the data with that from elsewhere (cf. Vikan et al. 2007). However, this project began as a comparative project, created to determine whether undergraduates in India and the UK had similar feelings about the environment and, perhaps, whether 'student culture' was uniform despite other cultural differences? The null hypothesis was that there should be no significant differences between the attitudes and values of these two student populations, despite their massively different cultures and life experiences.

Of course, life in Oxford, England, may not be as bucolic as it is painted. Student life and the drive for success still create stresses. However, the environment is not, in general, hazardous. Certainly, no-one is worried that they or their family are in immediate danger of being swept away by a landslide. So, whatever the cultural differences between undergraduate learners in Oxford and Uttarakhand, there is, inevitably, a huge difference in the intensity of their relationship with the environment as hazard. Table 4.2 compares two groups. First, it uses the combined data from the students of Uttarakhand. Second, it uses data from upper level learners reading Geography and/or Physical Geography and/or another discipline at Oxford Brookes University, UK. Table 4.2 again uses raw (unrecoded) scores but unlike Table 4.1 these are ranked according to the strength of the response. Of course, this means that the questions should segregate with odd-number NEP questions nearer the top and even-number NEP questions, the ones normally recoded, near the bottom. Once again, Table 4.2 compares the two populations to identify those questions where response patterns are significantly different.

In Table 4.2, these raw (unrecoded) NEP responses are ranked according to the strength of feeling expressed. The column of differences shows that, in almost every case, the views of the Uttarakhand respondents are held more intensely than those from England. This may be a matter of culture but more likely it reflects the emotional impact of the recent disaster. There is one lone exception, NEP8, where the English students strongly disagree that nature is strong enough to cope with the impacts of modern industrial nations, while those from more rural Uttarakhand are broadly neutral.

The rankings of the upper level Geography joint-honours undergraduates in England and the undergraduate Education and first year postgraduate GIS learners in Uttarakhand also differ significantly ($p = 0.0025$ in a Wilcoxon's matched pair's test). In most cases, both groups lean towards the same response and the differences, while often statistically significant, reflect only a matter of degree. However, there are a few areas of broad disagreement, where the two groups align on opposite sides of neutral. Here the Uttarakhand students believe that the eco-crisis idea has been much exaggerated (NEP10), that human ingenuity will keep the earth habitable (NEP4) and that humans will eventually learn enough to control nature (NEP14), while the English students do not. Additionally, the Uttarakhand students are unsure if nature is strong enough to cope with the impacts of modern industrial nations (NEP 8), while the English students think not. However, the Uttarakhand students do not believe that humans have any right to modify the natural environment for their own purposes, while the English students are unsure (NEP2).

In Table 4.2, the segregation of the positive and negative NEP questions is not complete, although 4/5 in the Uttarakhand top five are odd-numbered and 4/5 in the bottom five are even-numbered. The two exceptions are NEP11, which concerns the limits to growth, and which receives slight assent from both groups and NEP6,

Table 4.2 Comparison of Uttarakhand and English students using raw NEP scores (i.e. even numbered questions not recoded)

Uttarakhand University vs. English University	Uttarakhand mean (and rank)	Uttarakhand std. dev.	England mean (and rank)	England std. dev.	Difference of means	Kruskal-Wallis IS Test Sig
NEP7. Plants and animals have as much right as humans to exist.	4.44 (1)	.873	4.05 (1)	.852	0.39	<0.0005
NEP3. When humans interfere with nature it often produces disastrous consequences.	4.40 (2=)	.991	3.70 (7)	.819	0.70	<0.0005
NEP6. Earth has plenty of natural resources if we just learn how to develop them.	4.40 (2=)	.802	3.71 (6)	.772	0.69	<0.0005
NEP15. If things continue on their present course, we will soon experience a major ecological catastrophe.	4.30 (4)	1.155	3.79 (5)	.761	0.51	<0.0005
NEP5. Humans are severely abusing the environment.	4.28 (5)	.712	3.99 (2)	.644	0.29	<0.0005
NEP9. Despite our special abilities humans are still subject to the laws of nature.	4.03 (6)	1.112	3.97 (3)	.682	0.06	0.035
NEP13. The balance of nature is very delicate and easily upset.	4.01 (7)	.906	3.66 (8)	.816	0.37	<0.0005
NEP1. We are approaching the limit of the number of people the Earth can support.	4.00 (8)	1.075	3.85 (4)	.988	0.15	
NEP14. Humans will eventually learn enough about nature works to be able to control it.	**3.92 (9)**	.915	2.54 (14)	.735	1.38	<0.0005

(continued)

Table 4.2 (continued)

Uttarakhand University vs. English University	Uttarakhand mean (and rank)	Uttarakhand std. dev.	England mean (and rank)	England std. dev.	Difference of means	Kruskal-Wallis IS Test Sig
NEP4. Human ingenuity will insure that we do NOT make the Earth unlivable.	**3.63 (10)**	1.109	2.93 (11)	.821	0.70	0.008
NEP10. The so-called "ecological crisis" facing humankind has been greatly exaggerated.	**3.55 (11)**	1.203	2.77 (12)	.961	0.78	<0.0005
NEP11. The Earth is like a spaceship with very limited room and resources.	3.48 (12)	1.243	3.39 (9)	.846	0.09	
NEP8. Nature is strong enough to cope with the impacts of modern industrial nations.	3.09 (13)	1.340	**2.11 (15)**	.661	0.98	<0.0005
NEP2. Humans have the right to modify the natural environment to suit their needs.	**2.69 (14)**	1.248	3.04 (10)	.867	0.35	0.004
NEP12. Humans were meant to rule over the rest of nature.	2.43 (15)	1.213	2.58 (13)	.859	0.15	

Areas of disagreement are emboldened

which supports the belief that the Earth has sufficient resources if we just learn how to use them, which is strongly supported by both groups but, like many others, significantly more strongly by the Uttarakhand students ($p < 0.0005$).

Table 4.3 attempts to compare the results from Uttarakhand and England with other recent NEP surveys of undergraduate learners (Vikan et al. 2007; Denis and Pereira 2014; Ogunbode 2013). The means of both surveys lie close to but slightly below the overall mean, which is also low for those reported, mainly from the New World, by Schultz and Zelezny (1999). Only the Nigerian case study of students from Ibadan University has a significantly lower NEP mean (Ogunbode 2013). However, it is hard to know if this outcome has any meaning. Despite the number of studies from around the world and the large volume of data available, the comparison of NEP scores is a fraught business. Besides problems with the NEP itself, not least the fact that it originates from a dated American mind-set. There are problems

Table 4.3 Comparison of Uttarakhand and English students with other NEP survey results using a ranking of normally recoded NEP scores

Comparison by ranks with other NEP student surveys	Uttarakhand	England	Brazil	Norway	Turkey	Portugal	Romania	Nigeria	Means rank
NEP7. Plants and animals have as much right as humans to exist.	**1**	**1**	**4**	*9*	**1**	**3**	**1**	*9*	**3**
NEP3. When humans interfere with nature it often produces disastrous consequences.	**2**	*6*	*8*	*6*	**4**	**1**	**2**	**5**	**5**
NEP15. If things continue on their present course, we will soon experience a major ecological catastrophe.	**3**	**5**	**5**	**4**	**3**	**5**	*6*	**2**	**4**
NEP5. Humans are severely abusing the environment.	**4**	**2**	**1**	**2**	**2**	**2**	**3**	**3**	**1**
NEP9. Despite our special abilities humans are still subject to the laws of nature.	**5**	**3**	**3**	**1**	*6*	**4**	**5**	**1**	**2**
NEP13. The balance of nature is very delicate and easily upset.	*6*	*7*	*6*	**5**	**5**	*6=*	**4**	**4**	*6*
NEP1. We are approaching the limit of the number of people the earth can support.	*7*	**4**	15	14	*9*	*10*	*10*	*10*	*8*
NEP10. The so-called "ecological crisis" facing humankind has been greatly exaggerated.	**8**	**8**	**11**	**8**	**8**	**13**	**11=**	**5**	**7**
NEP11. The earth is like a spaceship with very limited room and resources.	*9*	*10*	*9*	13	11	11	*7*	11	*9*
NEP2. Humans have the right to modify the natural environment to suit their needs.	*10*	*9*	14	11	*7*	*9*	*8*	13	11
NEP12. Humans were meant to rule over the rest of nature.	11	11	*7*	*7*	*10*	*6=*	11=	14	12
NEP8. Nature is strong enough to cope with the impacts of modern industrial nations.	12	13	**2**	**3**	12	*8*	*9*	*6*	*10*
NEP14. Humans will eventually learn enough about nature works to be able to control it.	13	12	*10*	12	14	12	11=	12	13
NEP4. Human ingenuity will insure that we do NOT make the earth unlivable.	14	14	12	*10*	13	14	14	*7*	14
NEP6. Earth has plenty of natural resources if we just learn how to develop them.	15	15	13	15	15	15	15	15	15

due to the differences in the sample populations, differences in the modes of questionnaire distribution and caused by translation between languages. There are also problems due to culture, recent history and the fact that environmental attitudes evolve over time.

Nevertheless, despite these many caveats, Table 4.3 tries to compare the intensity of relative responses to (recoded) NEP items. Table 4.3, again, ranks these responses from pro-environmental to pro-anthropocentric. Hence, in Table 4.3, emboldened numbers indicate the greatest support for pro-environmental ideas, unemboldened numbers indicate support for anthropocentric notions, and italics show where there is no strong view or where pro-environmental and pro-anthropocentric views are balanced in the sample population. The distribution confirms that there is broad consistency amongst these different populations. The emboldened text dominates the top third of the table. The most obvious outliers among these are the support for the idea that Nature can cope with the impacts of the industrial nations (NEP8) among learners from Brazil and Norway, which also links to the notion that the Earth is not becoming overcrowded (NEP1).

By contrast, both the Uttarakhand and English surveys, as well as those from Turkey and Romania, feel most strongly positive about NEP7, which concerns ethics and the rights of animals and plants. In the case of Uttarakhand, this may well reflect local Hindu culture, which includes the notions of reincarnation and that the same soul exists in all creatures. In the case of England, it more likely reflects the course academic context; the questionnaire was administered as part of a course called 'The Ethical Geographer' (Boyd et al. 2008). However, overall, the most positive responses supported NEP5 and NEP9, which argue that humans are abusing the environment and are still subject to the laws of Nature.

There is also relative consensus at the opposite end of the scale where there are low scores for the idea that the earth has sufficient natural resources (NEP6), that human ingenuity will keep the Earth habitable (NEP4) and that human will eventually learn how to control Nature (NEP14).

The Uttarakhand students, after their ethical assertion (NEP7), most strongly assert their concern about environmental damage and human interference, the word disaster appears just once in the NEP and NEP3 is their second ranked assertion, where on average it is ranked fifth.

Collectively, the questionnaires have a consistent pattern and, despite their recent experience, the Uttarakhand questionnaire returns are not hugely different from the others. Indeed their most supported six questions are the same as the average and the same is true for those least supported.

In the Uttarakhand and England case studies, the NEP questionnaire was administered third in a battery of four. The second questionnaire included a set of four, essentially contradictory, statements about the learners' priorities in the world, whether self, family, Nation or environment, which nevertheless yielded important results (Table 4.4).

Two values are critical – the strength of the response and the rank. First, as before, the Uttarakhand scores are far higher than those from the students in England. Their reactions are far more intense. Second, the Uttarakhand students

Table 4.4 Priorities questionnaire

	Uttarakhand	Std. dev.	Uttarakhand rank	England	Std. dev.	England rank	Kuskal-Wallace IS Test
G1. In the final analysis, I put my own wellbeing first.	3.63	1.373	**4**	3.34	.925	**2**	0.039
G2. In the final analysis, I put my family's wellbeing first.	3.65	1.220	**3**	3.93	.831	**1**	0.021
G3. In the final analysis, I put my Nation's wellbeing first.	4.12	.977	**2**	2.59	.868	**4**	<0.0005
G4. In the final analysis, I put the environment's wellbeing first.	4.62	.822	**1**	3.16	.948	**3**	<0.0005

rank their environment's wellbeing as their priority, which is likely a consequence of the recent disaster. By contrast, the English students rank it third and express near neutral feelings.

The other major differences in the responses relate more to culture and politics. The English students put their family's wellbeing ahead of their own, where one might expect greater emphasis on their individual self. The same is true in the Uttarakhand sample but the difference is very small, while, given the character of Indian society, one might expect a much greater difference. After the importance of the environment, the greatest disagreement divides the priority given to the Nation. In a political climate of austerity and a pro-European part of English society, the state of the Nation is not prioritised. By contrast, in context of the cultural resurgence and rapid economic development of India, the Nation is prioritised above both self and family.

Some of the items from the other questionnaires also shed light on the ethics and values of the students (Table 4.5). Once again, the Uttarakhand responses show significantly greater intensity in most cases and the two 'most important' (Q7, Q8) questions yield useful results. In this case, the Uttarakhand learners prioritise the need for sustainability above concerns for human welfare, while the English learners rate them equal. Again, this may reflect concern about the recent disaster. The relatively high score awarded to the idea that humans and Nature are One (Q14) is, probably, a cultural difference; this belief is fundamental to Hindu tradition, where the English learners have only an academic exposure to such ideas through courses in environmental geography and ecology or Gaia Theory (Haigh 2014). However, both groups agree that a change in lifestyle is key to solving environmental problems (Q4) and, most hopefully for the future, both agree strongly that their generation has a duty of care to preserve the environment for future generations (Q6).

Table 4.5 Comparing attitudes of students from Uttarakhand and England: Supplementary questions

Additional items	Uttarakhand mean	Uttarakhand std. dev.	England mean	England std. dev.	Kruskal-Wallis IS Test Sig. Dif
Q6. Human beings have a duty to preserve the environment for future generations.	4.86	.463	4.43	.650	<0.0005
Q7. Environmental sustainability is the most important concern for human society.	4.55	.939	3.57	.880	<0.0005
Q4. Most environmental problems can be solved by changes in our life style.	4.29	.961	4.24	.622	0.008
Q14. 'Human beings and Nature are One' – a single living wholeness.	3.99	1.119	3.36	1.077	<0.0005
Q8. Human welfare is the most important concern for human society.	3.94	1.013	3.57	.756	<0.0005

5 Discussion

Vikan et al. (2007) compared NEP scores from 120 students in Brazil and the same from Norway. They recorded much higher scores among the Brazilians and attributed this to cultural differences. Here, for whatever reasons, although likely recent memories of the flood and landslide disaster, the Uttarakhand students' responses to these NEP questions are significantly more emphatic than those from the students in England.

The Uttarakhand learners feel strongly that human actions are responsible for environmental problems, such as those demonstrated by the recent landslide and flood disaster. However, they also have faith in the capacity of human ingenuity to resolve such problems, a view they do not share with their English counter-parts. Some of these key attitudes are stronger among female respondents from Uttarakhand. Rideout et al. (2010), among others, also found stronger endorsement among women although Vikan et al. (2007) found no differences in responses by gender.

In terms of Headwater management, a strong belief in the right to exist of other plants and animals (NEP7) as well as in the unity of Nature (Q14) are both resources that can be developed and used to promote environmental protection. Among Uttarakhand's learners, if not those in England, the wellbeing of the environment is a top priority, ahead of even Self and Family wellbeing (Table 4.4), which may be unusual in the context of a Hindu society. These attitudes are supported by the high priorities given to the welfare of future generations (Q6) and a willingness to resolve environmental problems through a change in lifestyle (Q4).

The NEP is the nearest thing there is to a global standard for exploring environmental attitudes (Dunlap 2008). However, a peer reviewer of this chapter advises that, while the NEP may be suitable for use at local or national level, it is unsuitable for comparison between countries, in the manner it has been used here and in the literature, because of differences in culture and also educational systems. Certainly, it is also dated and weak, where it comes to determining motivations and ethical values (cf. Schultz and Zelezny 1999). It has no space for sustainability or intergenerational justice and largely ignores spiritual matters. Unfortunately, this is just the tip of an iceberg of problems. In theory, the NEP contains five clusters of related questions called aspects. However, applying Varimax Factor Analysis to Dunlap et al. (2000)'s own case study discovered just four significant components, which explained 57% of the variance, while five factors in the Oxford Brookes data explained 55%, and seven in the Uttarakhand data explained 67% of the total variance. The first four components explained 47 and 45% respectively. Unfortunately, the poor definition of these 'aspects', whose items do not, in general correlate with each other and regularly correlate better with items from other 'aspects', means that their value is questionable. There are also much discussed questions about dimensionality explored by using Factor Analysis, a technique based on a correlation matrix that demands normally distributed data, which the NEP's Likert (i.e. ordinal) scale data is not (Dunlap et al. 2000; O'Connor 2010; Bernstein and Teng 1989)! There is also the problem that ordinal variables have no metric scale and a very restricted range, which is part of the correlation analysis problem (Marcus-Roberts and Roberts 1987). Even, when this issue is ignored, the derived factors provide little explanation and have little consistency in their loadings for different samples. In their case study, Principal Components Analysis finds all 15 items loading heavily (from .40 to .73) on the first (un-rotated) factor, which explains 31.3% of total variance. Here, in the data from England, it explains just 19.8% and from Uttarakhand only 15.0% of the variance.

However, one approach to the NEP questionnaire is neglected but, nevertheless, proves here to be both simple and beneficial. This is the technique of response ranking. Certainly, the ranking of Table 4.3 does demonstrate some of the consistencies in response patterns across studies that others have sought using more complex methods.

More generally, an analogy comes to mind. In the author's home area, South Wales, the valley roads tend to run north to south, a request for directions can be met with the response: "Sorry, you can't get there from here!" meaning that you should retrace your steps to another place, where you can cross into the next valley to continue your journey. This advice would also seem very appropriate in this case for this survey. For example, if the survey had been designed for administration in the aftermath of a disaster, instead of being printed ahead of time in England, it would have contained different, very different, questions.

Reflecting upon this issue, a peer reviewer of this chapter asks, innocently, why this study leans so heavily on statistical rather than qualitative, for example, interview data. Of course conducting interviews was fruitless, in this case, for several reasons. First was the problem of insufficient time. Second was the issue of language. More important was the problem of the questioner being a rather intimidating foreigner, arguably of 'high rank', which meant that questions are met with timid answers designed less to inform than to please and/or clear the path for a swift escape.

Perhaps most important, however, was the fact that all of these questions were creations of Western thinking and they embed, deeply, a Western sense of values and priorities. If they had been created from an Indian Hindu perspective, they would be very different (Tiwari and Misra 2012). This would be true both for Hindu interpretations of the causes of disaster and for questions that relate more generally to the state of the world and role of humans within it. Hindu spirituality connects religion with Nature, communities and their feelings for 'place' with, in this case, many levels of sacred space as well as of hazard (e.g. Mishra et al. 2010).

Some of these issues were addressed in communicating the findings about disaster causation from the contemporaneous landslide causation research project (Haigh and Rawat 2012). Here, landslide causes were diagnosed using the Hindu concept of the three sources of misery, the '*Tāpatraya*'. These three are: '*Ādhyātmika*', where the problem is self-inflicted; '*Adhibhautika*', where the problems are caused by the actions of other people and/or the material world of Nature, and '*Adhidaivika*', where the problems arise from unknown forces such as fate, chance; 'acts of God' as the insurance companies would have it (Pandya 1987; Jha 2011). A spiritual version of these definitions is found in that most widely read Hindu scripture, the *Srimad Bhagavad-gītā* 8, 1–4. In practice, the model fits the findings quite nicely.

In this environmental attitudes survey's results, there are hints that the Uttarakhand learners prioritise *adhibhautika* and accept ādhyātmika causes but there is nothing in the NEP that explicitly addresses the *adhidaivika*. This issue could have been resolved with a set of 'most important concern' statements (e.g. Q7, Q8, Table 4.4).

More generally, the Hindu worldview is underpinned by the '*Puruṣārtha*', or four goals of life. Three affirm life, they are: '*dharma*', which is correct or dutiful action, *artha*', success or prosperity, and '*kama*' which means enjoyment or love. The fourth is '*moksha*', liberation, which means release from the Hindu cycle or reincarnation through union with the divine. There is no Hindu rejection of either gathering wealth or enjoyment in this life, even if liberation from an endless cycle of reincarnation is won only by the renunciation of all worldly attachments.

The question remains how does the environment fit with this worldview? Is it mainly regarded as something requiring dutiful action, something providing kama as in the appreciation of Nature, or something that is a source of wealth. While Uttarakhand's learners reject the notion that humans have any rights over the environment; the Earth is respected as the mother goddess in Hindu religion. Again, many might regard environmental husbandry as an appropriate approach to dharma and artha. This question, again, could have been asked directly.

Finally, the path that leads to moksha has three steps, rather like the three steps of deep ecology and the three worldviews of Schultz (Haigh 2017). 'Triguṇa' Theory explains how the world is created by the interweaving interplay of three strands much as a colour image is created from RGB (red, green and blue) pixels. The three guṇa-s are: *Tamas*, which is everything static, dark, banal, heavy, obstructing, veiled; '*Rajas*', which is activity, energy, moving, changing; and '*Sattva*' which is everything: pure, light, sentient, serene, and peaceful. Sattva is the platform from which *moksha* may be achieved and Hindus are urged to overcome Tamas by positive action, *Rajas*, and so purify their lives to the state of *Sattva*, where action becomes unnecessary (Jacobsen 1999). Satish Kumar has applied these ideas in his vision of a spiritual compass for life (Kumar 2007). In Education, the first step is to dispel *Tamas*, literally, to remove ignorance, lethargy and indolence. This is achieved by engaging *Rajas* literally, engaging creativity, energy and passion, to create, construct, classify, and destroy. The third step is to rise above activity to *Sattva* and through reflection and compassion seek overview, synthesis and to replace the rajas discrimination of parts with an overview of wholeness (Haigh 2009, 2016).

These Hindu views about the fundamental qualities of Nature could have informed questions that gauge both perspectives on and reactions to environmental problems. Together, these ideas might provide the basis for a 'Hindu Environmental Paradigm' questionnaire that could produce answers more relevant and meaningful than the NEP. However, no matter how deep its roots in Hindu culture and theory, this customised questionnaire would be culture specific. It would not be appropriate for other groups in India, such as the Islamic or Christian communities or even, possibly, the more Westernised secular communities of India's largest cities.

Even so, this survey has highlighted some key aspects of environmental awareness that could be used as building blocks both for education and for headwater management. These include this community's concern for the rights of other plants and animals as well as for future generations, and both their concern about human abuse of the environment and its consequences and, unlike their English counterparts, a belief in the power of human ingenuity to make the situation better.

6 Conclusions

The Uttarakhand samples show that Geographers believe significantly more strongly than Education learners that humans are not meant to rule over the rest of Nature and remain subject to its laws while plants and animals have an equal right to exist. However, Education learners believe, significantly more strongly than Geographers, that the world is overcrowded, that the balance of Nature is easily upset by human actions, which progress will resolve. Meanwhile, Nature is strong enough to cope. Uttarakhand's women learners are significantly more complacent about the possibility of an eco-crisis than their male counterparts but also more certain that human interference is a root cause of disaster, although they also believe more strongly that technological progress will eventually provide a remedy.

In almost every case, the Uttarakhand student responses are significantly more extreme than those of their English counterparts. This may be a cultural difference but more likely it reflects the greater intensity of feelings following the recent disaster. The Uttarakhand students also believe significantly more strongly that human interference with nature leads to disaster and that, while the Earth has sufficient resources, the balance of nature is easily upset. The environment is suffering abuse and further problems lie ahead.

The English Geographers offer less support to anthropocentric notions. They disagree, significantly, with their Uttarakhand colleagues in not believing either that Nature is able to cope with the modern human impact or that humans will learn enough about the environment to control it. They do not believe that talk of 'ecocrisis' is exaggerated and worry more that human action will make the world uninhabitable.

Table 4.3 compares these two NEP surveys with other NEP student surveys using ranked data. The comparison reveals great consistency between responses ranked by intensity from positive to negative. Responses in the most strongly pro-environmental third of the questionnaire are nearly all at the top. Similarly, those at the bottom of these two NEP surveys are recoded anthropocentric items, very like those in the lowest ranks of other NEP surveys. However, at the top end, the Uttarakhand learners rank the rights of plants and animals and thoughts of disaster caused by human action higher than most other surveys.

The supplementary priorities questionnaire shows that Uttarakhand learners prioritise the environment above nation, family and self-interest, while English learners place the importance of both environment and nation below family and self-interest. The other supplementary questions indicate that the Uttarakhand learners accept, strongly, a duty of care for future generations. They agree also that lifestyle changes are key, prioritise sustainability above human welfare and they recognise the unity of Nature – all significantly more emphatically than their English counterparts, who had not experienced a recent disaster. Probably, most critical for headwater management is the Uttarakhand learners' ethical awareness of their own responsibilities, the rights of other creatures and future generations and their belief in the capacity of humans to remedy environmental ills. Of course, these findings defy the stereotypes of both populations, for example: Western learners are supposed to be less fatalistic and have a deep belief in the capacity of humans and technology to remedy the world's problems, while the reverse seems to be true.

Previously, many have questioned whether questionnaires like the NEP, which embed Western values and priorities, should be used in other-cultural situations? Certainly, if the NEP were created from a Hindu perspective, it would be very different. This chapter has considered some key resources for a Hindu 'NEP'-style questionnaire. They include: Hinduism's three sources of misery or Tāpatraya, the first three of the Puruṣārtha-s, the four goals of life, and the steps toward that fourth, moksha, suggested by 'Triguṇa' Theory with its three guṇa-s or modes of Nature. Of course, these Hindu views about the fundamental qualities of Nature could have informed the questions for a questionnaire that would have gauged Hindu perspectives on environmental problems much more effectively. They could form the basis of a 'Hindu Environmental Paradigm' questionnaire that might produce answers more relevant and meaningful than the NEP. Of course, the deeper its roots in Hindu culture, the more this customised questionnaire would be culture specific and, as such, inappropriate, even for other groups in India, such as the Islamic, Jain, Sikh or Christian communities or even, possibly, the more Westernised, more secular, communities of India's largest cities.

References

Anderson, B.A., Romani, J.H., Phillips, H., Wentzel, M. and Tlabela, K. (2007). Exploring environmental perceptions, behaviors and awareness: water and water pollution in South Africa. *Population and Environment*, 28(3), 133–161, https://doi.org/10.1007/s11111-007-0038-5.

Bernstein, I.H. and Teng, G. (1989). Factoring items and factoring scales are different: Spurious evidence for multidimensionality due to item categorization. *Psychological Bulletin*, 105, 467–477, https://doi.org/10.1037/0033-2909.105.3.467.

Bird, D.K. (2009). The use of questionnaires for acquiring information on public perception of natural hazards and risk mitigation – A review of current knowledge and practice. *Natural Hazards and Earth System Sciences*, 9(4), 1307–1325, https://doi.org/10.5194/nhess-9-1307-2009.

Boyd, W., Healey, R., Hardwick, S., Haigh, M., Klein, P., Doran, P., Trafford, J. and Bradbeer, J. (2008). 'None of Us Sets Out To Hurt People': The Ethical Geographer and Geography Curricula in Higher Education. *Journal of Geography in Higher Education*, 32(1), 37–50, https://doi.org/10.1080/03098260701731462.

Denis, H.D. and Pereira, L.N. (2014). Measuring the level of endorsement of the New Environmental Paradigm: A transnational study. *Dos Algarves: A Multidisciplinary e-Journal* 23, 4–26. http://www.dosalgarves.com/rev/N23/Completo23.pdf.

Dunlap, R.E. (2008). The new environmental paradigm scale: From marginality to worldwide use. *Journal of Environmental Education*, 40(1), 3–18, https://doi.org/10.3200/JOEE.40.1.3-18.

Dunlap, R.E. and Jones, R.E. (2002). Environmental concern: Conceptual and measurement issues. *In:* Dunlap, R.E. and Michelson, W. (Eds), *Handbook of Environmental Sociology*. Greenwood, London, pp. 482–524.

Dunlap, R.E. and Van Liere, K.D. (1978). The "new environmental paradigm". *Journal of Environmental Education*, 9(4), 10–19, https://doi.org/10.1080/00958964.1978.10801875.

Dunlap, R.E., Van Liere, K.D., Mertig, A.G. and Jones, R.E. (2000). New trends in measuring environmental attitudes: measuring endorsement of the new ecological paradigm: A revised NEP scale. *Journal of Social Issues*, 56(3), 425–442, https://doi.org/10.1111/0022-4537.00176.

Erdoğan, N. (2009). Testing the new ecological paradigm scale: Turkish case. *African Journal of Agricultural Research*, 4(10), 1023–1031, https://doi.org/10.5897/AJAR; http://www.academicjournals.org/journal/AJAR/article-abstract/40CFE5F35449.

Haigh, M. (2017). Connective practices in sustainability education. *Journal of Applied Technical and Educational Sciences ('jATES')*, 7(4), 6–30. [ISSN 2560-5429]. https://doi.org/10.24368/jates.v7i4.16. Open access from: http://jates.org/index.php/jatespath/issue/view/1/2017_4. (Accessed August 11, 2019).

Haigh, M. (2016). Using the Three Modes of Nature (Guṇa-s) in Invitational Education: Five levers for learning. *Other Education: The Journal of Educational Alternatives*, 5(1), 4–29, http://www.othereducation.org/index.php/OE/article/view/132/141.

Haigh, M. (2014). Gaia: 'Thinking like a Planet' as Transformative Learning. *Journal of Geography in Higher Education*, 38(1), 49–68, https://doi.org/10.1080/03098265.2012.763161.

Haigh, M. (2009). The Sattvic Curriculum: A 3-Level, Non-Western, superstructure for undergraduate education. *New Directions for Teaching & Learning*, 118, 61–70, https://doi.org/10.1002/tl.353.

Haigh, M. and Rawat, J.S. (2011). Landslide Causes: human Impacts on a Himalayan landslide swarm. *BELGEO: The Belgian Geographical Journal*, 11(3–4), 201–219 (Published July 2012). Available open access from http://belgeo.revues.org/6311 (accessed July 26th, 2013); http://journals.openedition.org/belgeo/6311 (accessed 4, August 2018); https://doi.org/10.4000/belgeo.6311.

Haigh, M. and Rawat, J.S. (2012). Landslide disasters: seeking causes: A case study from Uttarakhand, India, Chapter 18. *In:* Křeček, J., Haigh, M., Hofer, T. and Kubin, E. (eds), *Management of Mountain Watersheds*. Springer, Dordrecht, NL, pp. 218–253, https://doi.org/10.1007/978-94-007-2476-1_18.

Haigh, M.J., Rawat, J.S. and Bartarya, S.K. (1988). Environmental correlations of landslide frequency along new highways in the Himalaya: preliminary results. *Catena*, 15(6), 539–553, https://doi.org/10.1016/0341-8162(88)90005-7.

Hawcroft, L.J. and Milfont, T.L. (2010). The use (and abuse) of the new environmental paradigm scale over the last 30 years: A meta-analysis. *Journal of Environmental Psychology*, 30(2), 143–158, https://doi.org/10.1016/j.jenvp.2009.10.003.

Helm, S.V., Pollitt, A., Barnett, M.A., Curran, M.A. and Craig, Z.R. (2018). Differentiating environmental concern in the context of psychological adaption to climate change. *Global Environmental Change*, 48, 158–167, https://doi.org/10.1016/j.gloenvcha.2017.11.012.

Jacobsen, K.A. (1999). *Prakrti in Samkhya-Yoga: Material Principle, Religious Experience, Ethical Implications*. Peter Lang, New York.

Jha, H.B. (2011). The Economics of Peace: A Nepalese perspective. Observer Research Foundation (New Delhi), *Occasional paper*, 29, 1–30. Available from: https://www.orfonline.org/wp-content/uploads/2011/12/OccasionalPaper_29.pdf. (accessed: 01/05/2018).

Kumar, S. (2007). *Spiritual Compass: The Three Qualities of Life*. Totnes, UK, Green Books.

Lundmark, C. (2007). The new ecological paradigm revisited: Anchoring the NEP scale in environmental ethics. *Environmental Education Research*, 13(3), 329–347, https://doi.org/10.1080/13504620701430448.

Marcus-Roberts, H.M. and Roberts, F.S. (1987). Meaningless statistics. *Journal of Educational Statistics*, 12(4), 383–394, https://doi.org/10.3102/10769986012004383.

McIntyre, A. and Milfont, T.L. (2016). Who cares? Measuring environmental attitudes. *In:* Gifford, R. (ed), *Research Methods for Environmental Psychology*. UK, John Wiley & Sons, Chichester, pp. 93–114.

Mishra, S., Mazumdar, S. and Suar, D. (2010). Place attachment and flood preparedness. *Journal of Environmental Psychology*, 30(2), 187–197, https://doi.org/10.1016/j.jenvp.2009.11.005.

O'Connor, B.P. (2010). Cautions regarding Item-Level Factor Analyses. *Access*, 2010-02-22. Available from: https://people.ok.ubc.ca/brioconn/nfactors/nfactors.html. (accessed June 22, 2018).

Ogunbode, C.A. (2013). The NEP scale: Measuring ecological attitudes/worldviews in an African context. *Environment, Development and Sustainability*, 15(6), 1477–1494, https://doi.org/10.1007/s10668-013-9446-0; https://www.researchgate.net/publication/235955170_The_NEP_scale_Measuring_ecological_attitudesworldviews_in_an_African_context.

Pande, A. and Pande. L. (2020). Chapter 3: Community learning and resilience in the Himalayas, Uttarakhand, India. *In:* Pénzesné Kónya, E., Haigh, M. and Křeček, J. (eds). *Environmental Sustainability Education for a Changing World: Inspiration for and by Practitioners*. Springer, Cham, Switzerland & Capital Publishing Company, New Delhi.

Pandya, S.K.K. (1987) Hindu philosophy on pain: An outline. *In:* Brihaye, J., Loew, F., Pia, H.W. (eds), *Pain. Acta Neurochirurgica Supplementum*, 38. Springer, Vienna, pp. 136–146, https://doi.org/10.1007/978-3-7091-6975-9_25.

Pirages, D.C. and Ehrlich, P.R. (1974). *Ark II: Social Response to Environmental Imperatives*. W.H. Freeman, San Francisco.

Rideout, B.E., Hushen, K., McGinty, D., Perkins, S. and Tate, J. (2010). Endorsement of the New Ecological Paradigm in systematic and e-mail samples of college students. *Journal of Environmental Education*, 36(2), 15–23, https://doi.org/10.3200/JOEE.36.2.15-23.

Schultz, P.W. and Zelezny, L. (1999). Values as predictors of environmental attitudes: Evidence for consistency across 14 countries. *Journal of Environmental Psychology*, 19(3), 255–265, https://doi.org/10.1006/jevp.1999.0129.

Scolobig, A., Thompson, M. and Linnerooth-Bayer, J. (2016). Compromise not consensus: Designing a participatory process for landslide risk mitigation. *Natural Hazards*, 81(suppl. 1), 45–68, https://doi.org/10.1007/s11069-015-2078-y.

Sita Ramadas, Haigh, M. and Chauhan, S. (2014). Communicating sustainability within Britain's Hindu Community. *Sustainability*, 6(2), 718–740, https://doi.org/10.3390/su6020718; http://www.mdpi.com/2071-1050/6/2/718).

Tiwari, D.N. and Misra, A. (eds) (2012). *Environment Ethics: Indian Perspectives*. Varanasi, India: Banares Hindu University. Department of Religion and Philosophy. [ISBN 13: 978-81-924291-0-6].

Truelove, H.B. and Gillis, A.J. (2018). Perception of pro-environmental behavior. *Global Environmental Change*, 49, 175–185, https://doi.org/10.1016/j.gloenvcha.2018.02.009.

Vikan, A., Camino, C., Biaggio, A. and Nordvik, H. (2007). Endorsement of the new ecological paradigm: A comparison of two brazilian samples and one norwegian sample. *Environment and Behavior*, 39, 217–228.

Chapter 5
Understanding the Motivations, Perceived Benefits and Needs of Long-Term Volunteers on Environmental Projects

Marianne P. Kilmartin

1 Introduction

Practical environmental conservation depends heavily on the efforts of volunteers. However, relatively little research has sought to understand the motivations and perceived rewards of those engaged in environmental volunteering as distinct from those volunteering in other sectors. This study aims to explore the motivations and rewards perceived by a group of extreme long-term environmental volunteers. To date, very few studies have considered what encourages extreme long-term participation in environmental volunteering (West and Pateman 2016). This aims to explore the motivations and perceived rewards among a group of such extreme long-term environmental volunteers, who have returned to support the same environmental restoration project for as long as 28 years. Here, the question asked is how do the motivations or perceived rewards of these extreme volunteers differ from those who have supported the same project for shorter periods; in this case, less than 10 years?

Epidemiological studies detect long term (20-year) health and well-being-related benefits associated, specifically, with environmental volunteering (Pillemer et al. 2010); although, here, several volunteers have sustained persistent injuries through working in difficult conditions. It is already understood that community education is as important as self-education among the motivations of environmental volunteers (Measham and Barnett 2008; Merenlender et al. 2016). More generally, environmental volunteers perceive social benefits, including making friends and, sometimes, professional contacts through volunteering. The social aspect of volunteering is strongly related to enthusiasm intensity and ranks among the most important factors in retaining volunteers long term (Ryan et al. 2001; McDougle et al. 2011).

M. P. Kilmartin (✉)
Educational Advice and Guidance Department, Open University in the South, Oxford, UK

E. Pénzesné Kónya et al. (eds.), *Environmental Sustainability Education for a Changing World*, https://doi.org/10.1007/978-3-030-66384-1_5

2 Background of the Project

More than 78,000 ha of England and Wales have been affected by mineral workings and their waste tips. Many communities exist in landscapes that have been blighted by coal-mining and heavy industry with their younger generations leaving to seek a better life elsewhere. Today, the prosperity of the communities remaining in these post-industrial lands depends on attracting new industries but many employers are not keen to move into low-quality landscapes that are littered with industrial dereliction and waste tips. After the Aberfan disaster in 1966, official reclamation agencies began to 'tidy up' these damaged lands and place greater emphasis on environmental quality in post-industrial areas. Unfortunately, much of the land reclamation undertaken, especially for the large areas affected by opencast coal-mining, proved cosmetic and short-term because it was based on engineering concepts and emphasised structures (such as concrete gullies and weirs) and applied topsoils, which all need careful management and expensive maintenance (Haigh 2000). Seldom were the resources required for this kind of work available to affected local authorities and communities.

Here, these problems were complicated by natural weathering processes in the mine spoils. Here, the accelerated weathering of fragile minestones freshly exposed to the elements has flooded the soils with fine particles. These have become trapped in soil pores creating a high density and an almost entirely impermeable layer in the ground. Instead of infiltrating, more rainwater becomes runoff causing soil erosion, gullying, the over-loading of artificial drains and local flooding downstream. Once again, typically, the local communities affected by the degradation of so-called 'reclaimed land' do not have access to the funding required for further or new land reclamation works, so the task falls to voluntary sector initiatives of which there are few.

Since 1971, Haigh has undertaken research in what later became the UNESCO World Industrial Heritage area at Blaenavon, Torfaen, SE Wales (Haigh 1978). This research sought to monitor and understand the processes that caused land degradation on former opencast-coal-mine sites, especially compaction in the soil profile and the other factors that prevented the survival of vegetation cover. In 1991, land at Varteg, just south of Blaenavon, was acquired from Pontypool Parks Estates for research into the remediation of degraded, albeit formerly 'reclaimed', coal-land in the post-coal-mining environment.

This project sought solutions that, because of the lack of funding for the management, maintenance and repair of 'reclaimed' lands, would have to be self-sustaining, hence founded in biology, ecology and, perhaps, ecological engineering rather than conventional engineering. The strategy adopted, following ancient tradition, was forest fallowing, which means cultivating trees that, hopefully, will generate large quantities of organic matter, combat soil compaction by fostering the development of healthy soil crumbs, hold water and encourage water infiltration. This would encourage the evolution of a healthy soil ecosystem and so create and preserve soil fertility, which, in turn, would foster the development of a self-sustaining and self-improving geoecological system. This is now called the 'Cradle for Nature' strategy (Haigh 2018).

The main 'Cradle for Nature' project objective was to develop a reliable, inexpensive, self-sustaining, approach to restoring and revegetating lands that remained damaged after surface coal mining, which would be suitable for adoption by community-based land reclamation groups. Meanwhile, many technical problems had to be resolved included finding tree species able to withstand and better still thrive in the infertile, sterile, compacted, locally contaminated soil and hostile climatic conditions. Recent reports illustrate some of the progress the project has made towards this goal (Desai et al. 2018; Haigh et al. 2018, 2019; Plamping et al. 2017).

3 Funding and Staffing the Project

Funding for the project proved difficult to obtain. Part of the reason is that the degraded surface coal-mine lands selected for study are still, officially, counted 'reclaimed'. The UK Government has a category for unreclaimed land and for reclaimed land but not for reclaimed lands where the reclamation has failed, despite this being a large subset. Since the problem does not exist, officially, it has been impossible to gain funds from official agencies. However, photographic evidence proved sufficient to win seed-corn funding from Oxford Brookes University, then later, from 1988 until 1998, through a British Council sponsored link with the University of Forestry in Sofia, Bulgaria. An award from a small NGO (The Gaia Initiative of the Centre for International Peace Building) provided sufficient resources for the team to be able to fence a trial field of several hectares on formerly 'reclaimed' opencast-coal-mine disturbed land at Varteg on land loaned, at a peppercorn rent, by Pontypool Park Estates. Once this was done, the project applied to, and was accepted by, the Earthwatch programme, which places volunteers as lay citizen scientists on conservation and environmental monitoring projects around the world. Earthwatch provided its first two teams to the Varteg project in 1991. Over the next 13 years, Earthwatch placed around 400 volunteers from more than 30 nations into the project (Haigh 1998). These helping hands powered the project.

At this stage, the volunteers paid Earthwatch, so providing core-funding for both the Earthwatch organisation and for the field work. In addition, many volunteers had to fund their own travel expenses – often from far away countries. However, after 2004, the project continued without Earthwatch involvement through self-funding by those involved. Later, these volunteer teams were supplemented by academics, undergraduates and post-graduate students from the University of Wales Trinity St David, Swansea, who were supported by their university's staff development programme.

Typically, the annual pattern of fieldwork included an autumn tree-planting weekend and a longer summer camp for tree measuring, soil testing, and other related activities. Over time, many different stake-holders have invested in this project and its field-camps including private charitable foundations, commercial companies, community groups and visiting scientists. However without the support of its volunteers, especially the core group who returned to the project through many years, it would not have been possible to sustain the project's work across the three decades of its existence.

4 Motivations

It is said that two main aspects affect an individual's propensity to volunteer; 'internal' (personal) factors such as personality, attitude and social background, and 'external' situational factors such as finance, work and family commitments (Lockstone et al. 2002). This study focuses on the internal personal factors/motivations as it would be difficult to include all the situational factors that have affected the volunteers over such a long period. As well as these personal and situational factors, there is also the role of the project itself in retaining volunteers, which is also discussed.

Among many previous studies, Bruyere and Rappe (2007) surveyed 401 volunteers from six natural resources organisations in the USA. They found that, from the seven motivational factors that emerged: 'Helping the environment' came out as the most important, then, in decreasing order of importance: 'Learning', 'Social benefits' (i.e. meeting like-minded people, 'Values and Self-esteem', 'Project Organisation' and 'Career Benefits'. The final motivation that emerged was 'User', which related to people volunteering to improve an area where they live or work. The earlier 'Voluntary Functions Inventory Model' of Anderson and Cairncross (2005) proposed six motivators: 'Personal Values', 'Understanding'. 'Social', 'Career Benefits': 'Protective', which means overcoming the risk of negative thoughts about being better than others; and 'Personal Enhancement' (cf. building 'Self-esteem).

Previously, in this project's Earthwatch era, 140 'Cradle for Nature' project volunteers were asked to explain, first, why they wanted to join the project, as part of their application form, and, later, what were their most positive experiences, as part of their post-project evaluation (Table 5.1). Personal concerns for the environment were among the most mentioned reasons in the application but, curiously, did not feature at all in the post-project reviews. Instead, social factors, especially working with like-minded people, dominated along with a sense of achievement and feelings of belonging within the team. These same factors also feature strongly in the application forms. However, in post-project returns, the ranking of education sinks, 'local context' remains near the foot of the list, while 'enjoyment of outdoor activities' vanishes, possibly because of the weather. These rankings are rather different to those proposed by previous studies and the Anderson and Cairncross (2005) protective factor seems to be entirely absent, while social benefits, broadly described, are much more prominent among these early 'Cradle for Nature' Earthwatch volunteers. Previously, Clary et al. (1998) developed a classification of the individual benefits and needs that are fulfilled by volunteering and suggested that, when one or

Table 5.1 Previous survey of volunteer motivations and experiences on 'Cradle for Nature' project (Haigh 1998)

Application Form: Reasons for volunteering (number of mentions)	Post- volunteering project Evaluation Forms: Most positive experiences (number of mentions)
1. Interest in subject (environmental reconstruction/'healing the Earth') (42)	1. Enjoyed the company of like-minded/interesting people (41)
2. Build technical understanding – Self-education (38)	2. Working as part of a team (15)
3. Concern for the environment (generally) (30)	3. Sense of achievement (14)
4. Want the company of like-minded/interesting people (25)	4. Optimistic/shared enthusiasm for the project (12)
5. Want personal involvement in a 'worthwhile' activity (24)	5. Being a valued part of the project (11)
6. Work as part of a team (22)	6. Involvement with local community/heritage (9)
7. Enjoy outdoor activities (19)	7. Learning new things – increased technical understanding (9)
8. Commitment to local area of project (19)	*N.b. Most frequently listed negative experiences were: Bad weather (14), lack of time with the project (9) and assorted logistic and organisational issues (12)*
9. Learn new things – Useful for career (17)	*Table includes themes mentioned in more than 10% of responses.*
Sample size = 140; female: 91; male: 49; age: 20: 7; 20–40: 49, 40–60: 68, <60: 16; origins: UK: 97; USA: 24; W. Europe: 10. Rest of the world: 9.	*Sample size = 67; female: 40; male: 16; Don't know: 1. Origins: UK: 55; USA: 6, W. Europe: 2; Don't know: 4*

more of these benefits or social needs are met by participation in a project, then an individual is more likely to continue (Ryan et al. 2001: 645).

5 Collecting the Data

Structured interviews were carried out with 17 volunteers in 2017. These interviews were held sequentially across successive evenings at the project's field-base hotel in Wales. Each interview lasted around 30 min and was, with each respondent's express consent, recorded. Core questions included: how many 'Cradle for Nature' field camps had they attended, whether or not they did other environmental volunteering, why they had remained with this project for so long, what they felt was their special contribution to the project, why they volunteered for this project in the first instance, and what had been the long-term impact of their engagement with the project? Subsequently, transcripts of each of the recordings from these interviews were analysed to extract general themes and individual comments about what motivated the volunteers to persist with their volunteer engagement.

6 Results

Volunteer comments suggest that the motivations for volunteer involvement clustered around some broad themes, many of which, of course, have also been identified in previous studies. Here, 12 general motivation and benefit categories were identified and selected for further exploration and, again, many are shared with previous studies, albeit sometimes with slightly different aspects. They concern career development, enhancing self-image, gaining skills and knowledge, enjoyment of working outdoors, improving well-being, which includes a sense of feeling needed, ownership of the project and/or the trees planted, creating or sustaining friendships, helping the environment, helping the local community and developing their own self-identity as an environmentalist. Two further themes were also explored: one related specifically to further involvement in the project (which was included to help plan future work), while a second sought to find out whether long-term volunteers were also more likely to work with other environmental projects. The 'Protective' motivator of the Anderson and Cairncross (2005) 'Voluntary Functions Inventory Model', as in the previous survey (Haigh 1998), did not appear to be present at all and is not included.

Discussed below are the 12 principal motivational themes that were identified from the interviews and selected for further exploration (along with an extra question about whether the volunteers wished to continue working on the 'Cradle for Nature' project, again to help with future project planning).

1. *Identity*: Penner (2002) stressed the importance of how the volunteer is treated by the organisation. He said that the role identity of the volunteer should develop if the initial level of volunteering is to be maintained. The part that the organisation plays in whether the volunteers develop a role identity is discussed later. Here, the volunteers who had been coming to the project for a number of years had developed their own role identity in the team – often using skills and experience from their careers – such as the identification of worms and small mammals or data analyses. As one volunteer said *'I always felt I was bringing something and making things better'*. A long-term volunteer found the role of *'Bringing some organisation and consistence to the data collection and processing. I felt valued'*.
2. *Ownership*: Sklar and Ames (1985) and Sommer et al. (1994) found that there appeared to be a special attachment that comes with forestry volunteering and it is interesting to determine whether this is apparent from the beginning or whether this develops over time. Here, along with a sense of ownership, came a sense of pride and accomplishment particularly when the trees developed successfully (Haigh et al. 2018).
3. *Feeling that contribution is needed*: McDougle, Greenspan and Handy (2011) and Moskell, Broussard Allred and Ferenz (2010) mentioned that volunteering can make individuals feel needed. Some of the volunteers in these interviews indicated that they felt needed but it was not generally expressed in these words. Some felt that the project needed them – which it did. Some volunteers described

how their feelings of self-worth and self-confidence increased. The annual work camps were obviously important events in their lives and they seemed to really look forward to them. For some, the author felt, it was an opportunity to get away from every day responsibilities and do something for themselves. One of the long-term volunteers said it was '*Something just for me*'.

4. *Environmental benefits*: This is a motivation that is based on passionate support of **all** environmental issues. One of the most common comments in this study was '*I want to give something back … and help the environment*'. One volunteer described the reclamation work as '*watching the world come back*'. Both Křeček and Pažourková (2020) and Bruyere and Rappe (2007) found this motivational factor to be the strongest among the six or seven motivations (respectively) that they identify.
5. *Social benefits*: Studies by Anderson and Cairncross (2005) and Clary and Snyder (1999) identified the importance of the social aspects of volunteering. Some in this group looked forward to creating new friendships and some sought to sustain existing relationships. One long-term volunteer said '*I came really to meet everyone*' and had '*met incredible people*'. Křeček and Pažourková (2020) rank this motivation 3/6 in their study of Earthwatch volunteers in northern Bohemia, behind helping the environment and extending interdisciplinary knowledge. Bruyere and Rappe (2007) found this to be a second tier function in their study that emphasised the importance to volunteers of working with people with similar interests.
6. *Gaining skills and knowledge*: This is closely related to the career motivation (Theme 10) but involves a more general acquisition of technical skills such as tree planting, data collection, data analysis and land reclamation. Again, this motivation has been identified in many previous studies including: McDougle, Greenspan and Handy (2011), Handelman (2013) and Bruyere and Rappe (2007), who found 'learning' to be the second strongest motivation in their study group and thought that project work gave their volunteers an opportunity to learn about environmental issues from an insider perspective. In this Varteg study, nearly all of the volunteers mentioned that they had learned new skills. A few also mentioned that they welcomed the opportunity to compare notes with volunteers from other countries. As one new volunteer stated, it is about '*Just being out there and realising what can be done*'.
7. *Improved well being*: Anderson and Cairncross (2005) mention that some volunteers were hoping to overcome previous negative experiences – and develop renewed self-confidence and self-belief. Although this theme was addressed by a very few volunteers in these interviews, it had been a very powerful motivation for some of them. A relatively new volunteer said: '*My confidence grew each time – it gave me a way of doing something worthwhile.*' Another said that it was '*a life changing experience – more than just a benefit.*' A fairly new volunteer found that '*I suddenly learned that I had a passion for nature and sustainable development.*' As one long-term volunteer put it '*I wanted to see it through and help measure the trees that I planted*' and another said '*I believe in what we have done*'.

8. *Involved in other volunteering*: This theme was included to see if there was a correlation between the length of time they had volunteered at Varteg and their engagement in volunteering for other projects.
9. *Enhanced self image*: Anderson and Cairncross (2005) described this as the way in which the ego can develop in a positive way. This was mentioned by a number of the volunteers in the context of their interactions with friends as '*having something different to talk to them about*' and '*becoming known as a person who helps with environmental issues*'. It certainly seemed to make the individuals feel proud of their involvement. '*People think they know about land reclamation but they don't … so I tell them about it.*
10. *Benefits to career*: This was one of the more common motivations identified by previous, admittedly usually short-term, studies: Handelman (2013); McDougle, Greenspan and Handy, (2011); O'Brien, Townsend and Ebden (2010); Anderson and Cairncross (2005) and Clary et al. (1998). In this project, the career motivation manifested itself in two forms: For some volunteers, it involved gaining career-related experience for either a particular type of work or to use towards a qualification. For others (such as teachers), it related to CPD (Continuing Professional Development). For them, it was a way to learn more about opencast coal mining and the reclamation process to use in their teaching work. Previously, two of the volunteers interviewed had brought groups of school children on site and undertaken their own work camps at Varteg, while some others had taken the information learned about organising and running such field camps to use in work camps at home.
11. *Altruism*: Referred to by various different names in previous studies (such as 'values') this relates to a volunteer giving their time and (often) money to support a specific cause that they believe in. Davis Smith (1999) developed a broad typology of altruistic motivations, which included providing a service to others– maybe improving the environment for the whole community. Some of the interviewees in this Varteg study referred to other environmental volunteering that they did – which frequently improved the environment for the community, while one volunteer discovered '*A passion for people living better lives, natural lives closer to nature*'. Bruyere and Rappe (2007) suggest a 'user' motivation, which relates to the idea that people volunteer to work in and improve an area that they use themselves. However, most of the Varteg volunteers didn't live in the immediate area. Some came here 'on holiday', with the work camp often counting as a holiday activity. However, the area is very far from being a conventional holiday destination, especially in the years before the Blaenavon World Heritage Industrial Area was recognised. In the Varteg case, the volunteers' personal attachments to the area tended to be small and distant, while the material benefits they might gain from its restoration were smaller still. So, the motivation here is closer to pure, self-less, altruism, at least as far as the local area is concerned.
12. *Working outdoors*: Again this was counted a common motivation in both previous studies and the earlier times of this Varteg 'Cradle for Nature' project (Haigh 2016; Kragg 2017). In the USA, an open ended question about the most

important reason for volunteering, yielded ‘getting outside’ as the third most mentioned response (Bruyere and Rappe 2007). Despite the often inclement Welsh weather, many volunteers said that they enjoyed working in the fresh air with lovely views – often comparing it with their normal desk jobs and several commenting wryly about coping with horizontal rain and low temperatures during late November tree planting sessions!

The final extra question explored the volunteers’ enthusiasm to continue. This was added because the analysis as tree planting and the bulk of the programme of tree and soil measuring was approaching its natural conclusion, at least on this site’s original test fields. Twenty-year growth and vitality data had been collected for all the trees planted on the original test field and there was no plan to collect 30-year data. So, while much other work remained to be done, this question aimed to gauge the volunteers’ enthusiasm for continuing their work into the 2020s.

Data Analysis and Further Findings

The first step in this study was to collect the motivations and benefits that each individual had mentioned, code them to each theme, and rank these data by the frequency with which each theme was mentioned. The dispersion of this data suggests a natural separation in the data set which divides those who have been volunteers for less than 10 years (<10 years) and those who have been volunteers for more than 15 years (>15 years).

This separation is apparent in Fig. 5.1, which graphs the number of years each volunteer has contributed to the project against the number of reasons under different themes that they suggested in support of their volunteer participation. This simple, linear regression comparing ‘Years vs. Reasons’ for the volunteers is highly statistically significant ($p < 0.005$). It shows that the longer volunteers remain with the project, the more reasons they are able to describe for so doing.

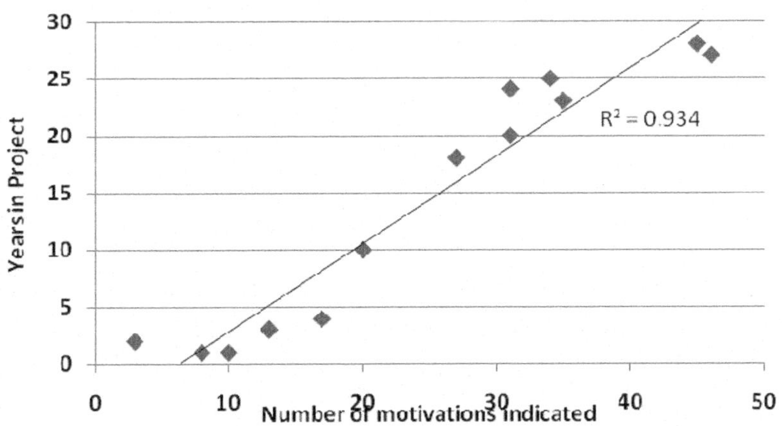

Fig. 5.1 Regression: Years vs. Reasons

However, the separation, which is apparent in the data displayed by Fig. 5.1, may also be used to compare the responses of shorter-term, although often quite long-term volunteers (<10 years), with the project's core of extreme long-term volunteers (>15 years), whose commitment typically exceeds 20 years.

The next step was to determine how the reasons/motivations of the longer term volunteers (>15 years) differ from those of the shorter term group (<10 years). Figure 5.2 compares counts of the mention of a particular motivation by individual volunteers. The table is constructed using the ranking of motivations most commonly listed by volunteers who had been with the project for more than 15 years.

Figure 5.2 confirms that volunteers in both categories had similar reasons for working in this environmental project but the longer term volunteers had almost twice as many as reasons for remaining with the project than the shorter term volunteers. Figure 5.2 shows that the >15 years volunteer group were: seven times more likely than the <10 years volunteers to feel that the project was part of their identity; six times more likely to feel a sense of ownership; more than twice as likely to want to continue with this Cradle for Nature project and twice as likely to enjoy the social benefits, to feel that their contribution is needed and that they were making a positive contribution to the environment.

It also shows that there are only three areas where the shorter term volunteers emphasise a particular motivation more than the longer term volunteers. These are: 1. value for the c.v., perhaps because the longer term volunteers are older, often

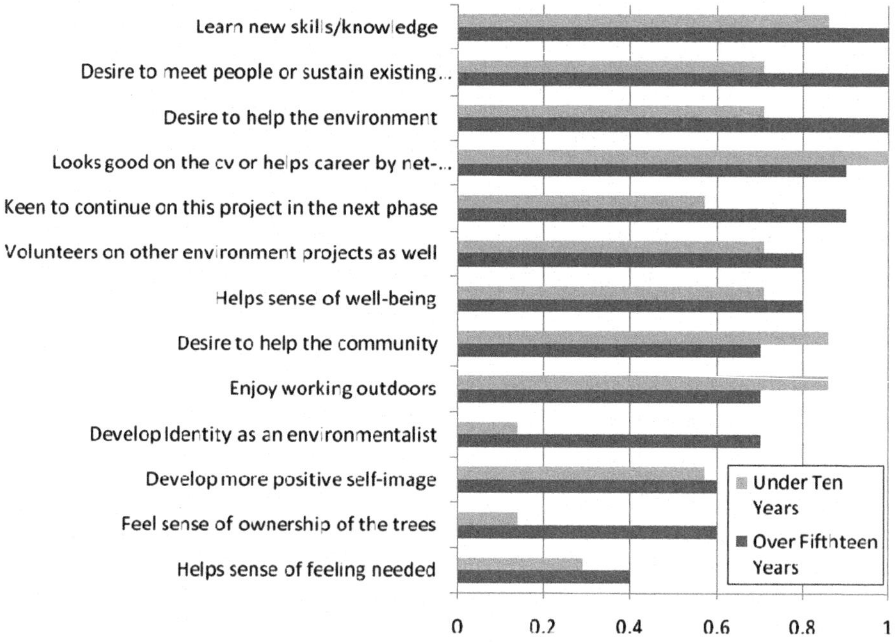

Fig. 5.2 Motivation: Comparing frequencies of motives mentioned by volunteers with >15 years engagement ($n = 101$) with <10 years ($n = 51$)

retired, so career is no longer an issue; 2. the desire to help the local community, which suggests that the longer term volunteers have relatively greater intrinsic motivations; and 3. the enjoyment of working outdoors, perhaps because, for the longer term volunteers, this has less novelty and, indeed, become more of a chore and a challenge for these older people.

Table 5.2 displays the differences in the ranked counts of motivations between the two categories of volunteers and so emphasises the differences between these two viewpoints. The greatest differences are the desire to develop an environmentalist identity, which is greater in the >15 years volunteers, as are the sense of ownership and wish to continue with the project. A desire to help the environment and both sustain existing friendships and meet new people, which were also more often

Table 5.2 Ranking of differences in the ranked counts of motivations between >15 years and < 10 years volunteers

Motivation (benefits)	Over 15 years	Ranking	Under 10 years	Ranking	Difference between <10/>15 years	Ranking of difference
Develop identity as an environmentalist	0.7	8=	0.14	12=	*0.56*	*1*
Feel sense of ownership of the trees	0.6	11=	0.14	12=	*0.46*	2
Keen to continue on this project in the next phase	0.9	4=	0.57	9=	*0.33*	*3*
Desire to help the environment	1	1=	0.71	5=	*0.29*	*4=*
Desire to meet people or sustain existing friendships	1	1=	0.71	5=	0.29	4=
Learn new skills/ knowledge	1	1=	0.86	2=	*0.14*	*6*
Helps sense of feeling needed	0.4	13	0.29	11	*0.11*	*7*
Helps sense of well-being	0.8	6=	0.71	5=	*0.09*	*8=*
Volunteers on other environment projects as well	0.8	6=	0.71	5=	*0.09*	*8=*
Develop more positive self-image	0.6	11=	0.57	9=	*0.03*	*10*
Enjoy working outdoors	0.7	8=	0.86	2=	*−0.16*	*12=*
Desire to help the community	0.7	8=	0.86	2=	*−0.16*	*12=*
Looks good on the cv or helps career by net-working	0.9	4=	1	1	*−0.01*	*14*

mentioned by the >15 years volunteers. The smallest differences involved the wish for career development, to help the community, to work outdoors, to develop a better self-image and sense of well-being and to feel needed as well as personal involvement with other projects.

7 Discussion

As an interviewer, it was a both humbling and inspirational experience talking to volunteers who give up so much of their time over so many years and sometimes travelled so far to brave the hawthorns and often inclement weather at Varteg. Their constancy, persistence and above all loyalty to the 'Cradle for Nature' project and their fellow team members was hugely impressive. A report from The Conservation Trust charity, which compares, short term and 'long term' volunteers (by which they mean those with more than three years(!) of engagement) also finds that the longer term volunteers travel greater distances and commit more of their personal time to the project (Seymour and Haklay 2017). Seymour and Haklay (2017) also describe, but do not explore, a similar process of distillation by which the many initial volunteers reduce to a handful of long-term volunteers, here also less than 2% of the total.

The Effects of Good Team Management

How an organisation treats their volunteers influences their level of enthusiasm and their likelihood of continuing to support a project (Penner 2002). The volunteers in the Varteg project appeared to appreciate the way the 'Cradle for Nature' was managed and, obviously, felt a close bond with its team leader. Of course, managing a team that includes a wide range of individuals from different backgrounds is challenging as their motivations will vary considerably. This is why determining motivation is such an important issue for the manager of this sort of NGO team. In day to day practice, it is important that the scale of the investment expected of each individual must be matched to their degree of personal motivation (Haigh 1998). If too much is asked, or indeed too little, of a volunteer, they are likely to respond negatively.

Bruyere and Rappe (2007) observe that volunteers want to be part of a well organised project, know what is expected of them and work with a committed leader. They found that a well organised project would usually accomplish more and that participants felt that their free time was being used wisely. They also found that participation in project decision making was important to the volunteers, as were activities that campaigned or raised awareness about the conservation issues. Here, some volunteers elected to take over project management duties including: hotel booking, data curation and coding, field survey, imaging, and liaison with local communities. Some of the Varteg volunteers elected to work closely with the local community and local conservation groups, with special teams constructed to undertake such activities; the volunteers felt it was 'their' woodland and so helped both to protect it and publicise the work being done.

Measham and Barnett (2008) argue that environmental programmes which allow volunteers to pursue their interests, increase social contacts, and feel like they are making a difference are more likely to be successful in the long term. In the Varteg 'Cradle for Nature' project, volunteers were encouraged and enabled to develop their own interests and bring their personal special skills to the project. To date, volunteer-led sub-projects have explored aspects of biodiversity (floristic change, soil microbiology, earthworms, small mammal populations), as well as the development of tree root systems, seed-fall and so on.

West and Pateman (2016) suggest that a system of rewards is a good way of showing project volunteers that they are valued and also a way of motivating them to continue their involvement; however, the research evidence on this is sparse (Locke et al. 2003). A UK study of environmental volunteering suggested that there were several ways of motivating and rewarding volunteers but also that some proved counterproductive (O'Brien et al. 2010). However the list includes thanking volunteers (verbally and/or through newsletters), offering award schemes and social events, and, less usefully, providing incentives such as long service badges, which could easily prove divisive, and/or volunteer discount cards for stores or cafes, which might seem to trivialise the volunteers efforts, although, as Earthwatch insists, providing some small souvenir of the camp may be welcome (O'Brien et al. 2010). Here, at Varteg, as well as receiving mailings, souvenirs and 'thanks', most long-term volunteers were also encouraged to act as co-authors of the projects major scientific publications and, where possible, take the role of lead author (e.g. Plamping et al. 2009, 2017). Some found this a very surprising and important acknowledgement of the work they had done. As one volunteer said 'It's amazing to see your name on an (academic) paper'.

In the earlier study of Varteg volunteer motivations, Haigh (1998) suggested three key issues to be taken into account in project design and management. First, the volunteers should not feel that their contribution to the project is greater than they wished to contribute so the project does not become a burden. Similarly, they should not be allowed to feel overloaded or stressed. If this happens, of course, they will back away. However, they may be pleased to find that the project's work is 'demanding' and 'a challenge'. Second, the volunteers should feel that the project is of benefit to them. If it does not meet their personal goals, they will probably shift their focus towards an activity that does. It is important to remember that many of the shorter term volunteers value the perceived benefits to their Curriculum Vitae and personal development. Finally, the volunteers must believe that what they are doing is worthwhile. If they lose their faith in the project's direction and purpose, if they begin to feel that their personal contribution is not important, their support for the team and its work will, inevitably, fade away. Here, while the Varteg volunteers also receive regular bulletins on progress and messages of thanks, the 'Cradle for Nature' project's successes (and occasional failures) are highly visible in the form of the growth and survival of the newly planted forest and/or its increasing biodiversity and also its contributions to the scientific literature. These describe how well the trees have grown under different treatments and the ways in which these new trees have regenerated a healthier soil eco-system, encouraging earthworms, small mammals, etc. Of course, the volunteers are deeply aware (and frequently reassured) of the fact that none of this could have been achieved or documented without their volunteered inputs, which are recognised in the publications that bear their names as co-authors of the key project findings (e.g. Haigh et al. 2018, 2019; Plamping et al. 2009, 2017).

Cultural and Gender Differences

Sloane and Pröbstl-Haider (2019) collected empirical and quantitative data from a total of 473 environmental volunteers via a standardised online questionnaire: 255 from Austria and 248 from Great Britain. There were significant differences between the two populations including their educational backgrounds, when volunteering started and for how long it was sustained. However, the lead finding was that the British volunteers were strongly motivated by career opportunities and the learning experience, while the Austrian volunteers were mainly attracted by value and esteem. These differences are attributed to differences in the cultural values between the two countries.

In the USA, Taniguchi (2006, p. 91) reports that 'women are significantly more likely than men to volunteer'. This is certainly the case with the Varteg project where, as in the USA, work commitments and domestic responsibilities play an important role in each individual's capacity to volunteer.

In New Zealand, Fukushima (2016), undertook surveys of volunteer characteristics and motivations at Hauraki Gulf eco-leisure/tourism destinations and collected a total of 396 valid responses. This study explores the differences between 'episodic' and 'on-going' volunteering, which could, in some ways, be similar to the Varteg 'long term' and 'short term' model. Exploratory Factor Analysis suggested six factors of motivation, most of which had also been identified by previous studies. They were *values of nature, career, enhancement, protective, social,* and *leisure*. 'On going' volunteers put more importance on the exogenous factors such as *values of nature* and *leisure*, while 'episodic' volunteers were more motivated by endogenous factors such as *enhancement* and *protective* (Anderson and Cairncross 2005), which seem absent from the Varteg group. Fukushima (2016) also found that *place attachment* was stronger in the 'on-going' group. This is also reflected in the Varteg >15-year group who identified more strongly and felt a greater sense of ownership with the project site.

Differences Between <10-Year and >15-Year Volunteers

Results show that the >15 years volunteer group were: seven times more likely than the <10 years volunteers to feel that the project was part of their identity. This confirms the findings from previous studies, which show that a sense of ownership develops over time. However, there is no guarantee that this difference occurs around the 10-year point and further work would need to be done to confirm the actual timing. Additionally, the Varteg 'Cradle for Nature' project volunteers were six times more likely to feel a sense of ownership; more than twice as likely to want to continue with the project's work and twice as likely to enjoy the social benefits, to feel that their contribution is needed, and that they were making a positive contribution to the environment. Here, both categories had similar reasons for working in this environmental project but the longer term volunteers listed almost twice as many as reasons for remaining with the project than the shorter term volunteers. They were also more likely to be involved with other environmental volunteering projects, perhaps because, being older in the main, they had more time to build up contacts or because they had fewer home and work commitments.

These 'Cradle for Nature' project results also show that there are just three areas where the shorter term volunteers emphasise a particular motivation more than the longer term volunteers. These are: 1. value for enhancing their Curriculum Vitae or Resume, perhaps because the longer term volunteers are older, often retired, so career is no longer an issue; 2. the desire to help the local community, which suggests that the longer term volunteers have relatively greater intrinsic motivations; and 3. the enjoyment of working outdoors, something that also becomes tempered with age given the persistently challenging working conditions on the Varteg.

8 Conclusion

This study finds that:

1. The longer term, >15-year, volunteers felt a greater sense of ownership of, and identity with, their project than those who had <10 years participation. They were also twice as likely to feel that they were making a positive contribution to the environment, that their contribution is needed, and to enjoy the social benefits.
2. The social side of the volunteering was important to both groups but more so to the long-term group. The relationships that they had developed were more important to them. However, despite the fact that some of them were meeting up outside the project, there appeared to be a general feeling that they only wanted to continue coming to the Varteg work-camps if there was genuine environmental work to do. Their relationships seemed to be based on this shared work experience.
3. Individuals who had been supporting this project for such a long time were more likely to be involved with other environmental volunteering projects.

References

Anderson, E. and Cairncross, G. (2005). Understanding and managing volunteer motivation: Two regional tourism cases. *Australian Journal on Volunteering*, 10(2), 7–17.

Bruyere, B. and Rappe, S. (2007). Identifying the motivations of environmental volunteers. *Journal of Environmental Planning and Management*, 50(4), 503–516, https://doi.org/10.1080/09640560701402034

Clary, E.G. and Snyder, M. (1999). The motivations to volunteer: Theoretical and practical considerations. *Current Directions in Psychological Science*, 8(5), 156–159, https://doi.org/10.1111/1467-8721.00037.

Clary, E.G., Snyder, M., Ridge, R.D., Copeland, J., Stukas, A.A., Haugan, J. and Miene, P. (1998). Understanding and assessing the motivations of volunteers: A functional approach. *Journal of Personality and Social Psychology*, 74(6), 1516–1530. https://doi.org/10.1037/0022-3514.74.6.1516.

Davis Smith, J. (1999). *Volunteering and Social Development.* A background paper for discussion at an expert group meeting. New York, November 1999. Institute for Volunteering Research, London.

Desai, M., Haigh, M. and Walkington, H. (2018). Phytoremediation: metal decontamination of soils after the sequential forestation of former opencast coal land. *Science of the Total Environment*, 656, 670–680, https://doi.org/10.1016/j.scitotenv.2018.11.327.

Fukushima, N. (2016). Environmental volunteer experiences at eco-leisure destinations in New Zealand. Auckland University of Technology, Masters thesis, https://core.ac.uk/download/pdf/74310967.pdf and https://openrepository.aut.ac.nz/handle/10292/10191.

Haigh, M. (2019). Slow science: Benefits for the management of headwater catchments. *Land Use Policy*, 80, 449–455, https://doi.org/10.1016/j.landusepol.2017.12.041.

Haigh, M. (2018). Building a 'Cradle for Nature': A paradigm for environmental reconstruction. *In:* Chapter 32, Prasad, M.N.V., de Campos, P.J. and Maiti, S.K. (eds), *Bio-Geotechnologies for Mine Site Rehabilitation.* Amsterdam, Elsevier, pp. 593–616, https://doi.org/10.1016/B978-0-12-812986-9.00032-4

Haigh, M. (2016). Fostering global citizenship – Tree planting as a connective practice. *Journal of Geography in Higher Education*, 40(4), 509–530, https://doi.org/10.1080/03098265.2016.1150438.

Haigh, M. (2000). The aims of land reclamation. *In:* Haigh, M. (ed), *Reclaimed Land, Erosion Control, Soils and Ecology.* AA Balkema (Taylor & Francis, 2012), Land Reconstruction & Management 1, Rotterdam, pp. 1–20.

Haigh, M. (1998). Working with NGO Environmentalism for land reclamation research. *In:* Moxen, J. and Strachan, P.A. (eds), *Managing Green Teams: Environmental Change in Organisations and Networks.* Sheffield, Greenleaf, p. 69, https://doi.org/10.9774/GLEAF.978-1-909493-02-5_6.

Haigh, M. (1978). *Evolution of Slopes on Artificial Landforms.* Blaenavon, UK. Chicago, USA, University of Chicago (Geography Research Paper 183).

Haigh, M., Desai, M., Cullis, M., D'Aucourt, M., Sansom, B., Wilding, G., Alun, E., Garate, S., Hatton, L., Kilmartin, M., Panhuis, W. and Jenkins, R. (2019). Composted municipal green waste enhances tree success in opencast coal-land reclamation in Wales. *Air, Soil and Water Research*, 12(1), 1–10, https://doi.org/10.1177/1178622119877837.

Haigh, M., Reed, H., D'Aucourt, M., Flege, A., Cullis, M., Davis, S., Farrugia, F., Gentcheva-Kostadinova, S., Zheleva, E., Hatton, L., Plamping, K., Powell, S., Panhuis, W., Sansom, B., Sawyer, S., Wilding, G. and Woodruffe, P. (2018). Effects of initial fertiliser treatment on the 10-year growth of mixed woodland on compacted opencast coal-mine spoils, S. Wales. *Land Degradation and Development*, 29, 3456–3468, https://doi.org/10.1002/ldr.3111.

Handelman, C. (2013). Natural area studentship volunteers: motivations, attitudes, behaviours. Portland State University, Dissertations and Theses Papers 1058, https://doi.org/10.15760/etd.1058.

Kragg, G. (2017). A holistic approach to environmental volunteering: connections between motivation, well-being and conservation achievement. University of Bournemouth, UK, PhD thesis. http://eprints.bournemouth.ac.uk/29397/2/KRAGH%2C%20Gitte_Ph.D._2017.pdf.

Křeček, J. and Pažourková, E. (2020). Chapter 6. Education strategy supporting environmental services of mountain forest catchments. *In:* Pénzesné Kónya, E., Haigh, M. and Křeček, J. (eds), *Environmental Sustainability Education for a Changing World: Inspiration for and by Practitioners.* Springer, Cham, Switzerland & Capital Publishing Company, New Delhi.

Locke, M., Ellis, A. and Davis Smith, J. (2003). Hold on to what you've got: The volunteer retention literature. *Voluntary Action*, 5(3), 81–89, http://www.ivr.org.uk/VA+Documents%2fVA5_3%2farticle5_lockeetal.pdf.

Lockstone, L., Jago, L. and Deery, M. (2002). The propensity to volunteer: The development of a conceptual model. *In:* CAUTHE 2002: Tourism and Hospitality on the Edge; Proceedings. Edith Cowan University Press, Perth, W. Australia, pp. 377–390.

McDougle, L.M., Greenspan, I. and Handy, F. (2011). Generation green: Understanding the motivations and mechanisms influencing young adults' environmental volunteering. *International Journal of Nonprofit and Voluntary Sector Marketing*, 16(4), 325–341, https://doi.org/10.1002/nvsm.431.

Measham, T.G. and Barnett, G.B. (2008). Environmental volunteering: Motivations, modes and outcomes. *Australian Geographer*, 39(4), 537–552.

Merenlender, A.M., Crall, A.W., Drill, S., Prysby, M. and Ballard, H. (2016). Evaluating environmental education, citizen science, and stewardship through naturalist programs. *Conservation Biology*, 30(6), 1255–1265, https://doi.org/10.1111/cobi.12737.

Moskell, C., Broussard Allred, S. and Forenz, G. (2010). Examining volunteer motivations and recruitment strategies for engagement in urban forestry. *Cities and the Environment (CATE)*, 3(1), article 9, https://digitalcommons.lmu.edu/cate/vol3/iss1/9.

O'Brien, L., Townsend, M. and Ebden, M. (2010). 'Doing something positive': Volunteers' experiences of the well-being benefits derived from practical conservation activities in nature. *Voluntas: International Journal of Voluntary and Nonprofit Organizations*, 21(4), 525–545.

Penner, L.A. (2002). Dispositional and organizational influences on sustained volunteerism: An interactionist perspective. *Journal of Social Issues*, 58(3), 447–467, https://doi.org/10.1111/1540-4560.00270.

Pillemer, K., Fuller-Rowell, T.E., Reid, M.C. and Wells, N.M. (2010). Environmental volunteering and health outcomes over a 20-year period. *The Gerontologist*, 50(5), 594–602, https://doi.org/10.1093/geront/gnq007.

Plamping, K., Haigh, M., Reed, H., Woodruffe, P., Fitzpatrick, S., Farrugia, F., D'Aucourt, M., Flege, A., Sawyer, S., Panhuis, W., Wilding, G., Cullis, M. and Powell, S. (2017). Effects of initial planting method on the performance of mixed plantings of alder and oak on compacted opencast coal-spoils, Wales: 10-year results. *International Journal of Mining, Reclamation and Environment*, 31(4), 286–300, https://doi.org/10.1080/17480930.2016.1139528.

Plamping, K., Haigh, M., Cullis M. and Jenkins, R.E. (2009). Evaluation of Cambial Electrical Resistance (CER) for the appraisal of tree vitality on reclaimed coal lands. *International Journal of Mining, Reclamation and Environment*, 23(1), 21–32, https://doi.org/10.1080/17480930801999419.

Ryan, R.L., Kaplan, R. and Grese, R.E. (2001). Predicting volunteer commitment in environmental stewardship programmes. *Journal of Environmental Planning and Management*, 44(5), 629–648, https://doi.org/10.1080/09640560120079948.

Seymour, V. and Haklay, M. (2017). Exploring engagement characteristics and behaviours of environmental volunteers. *Citizen Science: Theory and Practice*, 2(1), art. 5, 1–13, https://doi.org/10.5334/cstp.66.

Sklar, F. and Ames, R.G. (1985). Staying alive: Street tree survival in the inner-city 1. *Journal of Urban Affairs*, 7(1), 55–66, https://doi.org/10.1111/j.1467-9906.1985.tb00077.x.

Sloane, G.M.T. and Pröbstl-Haider, U. (2019). Motivation for environmental volunteering – A comparison between Austria and Great Britain. *Journal of Outdoor Recreation and Tourism*, 25, 158–168, https://doi.org/10.1016/j.jort.2019.01.002.

Sommer, R., Leary, F., Summit, J. and Tirrell, M. (1994). The social benefits of resident involvement in street-tree planting. *Journal of Arboriculture*, 20(3), 170–175.

Taniguchi, H. 2006. Men's and women's volunteering: Gender differences in the effects of employment and family characteristics. *Nonprofit and Voluntary Sector Quarterly*, 35(1): 83–101, https://doi.org/10.1177/0899764005282481.

West, S. and Pateman, R. (2016). Recruiting and retaining participants in Citizen Science: What can be learned from the volunteering literature? *Citizen Science: Theory and Practice*, 1(2), art. 15, 1–10, https://doi.org/10.5334/cstp.8.

Chapter 6
Education Strategy Supporting Environmental Services of Mountain Forest Catchments

Josef Křeček and Eva Pažourková

1 Introduction

The concept of environmental services is based on the link between flows of values to human societies as a result of the state and quantity of the natural capital (MEA 2005). Mountain watersheds have played a critical role in maintaining a sustainable flow of mountain resources to extensive lowland areas. Particularly, their role as global "water towers" is supposed crucial to the welfare of humankind (Viviroli et al. 2007). For a long time, human society has contributed little to a stewardship or renewal of mountain watersheds (FAO 2006). Thus, in a global scale, the evidence of water quality deterioration is attributed mainly to the poor management of water resources recharge (Biswas et al. 2014). Mountain catchments face a large variety of degradational processes; natural assets are flowing downhill at unsustainable rates with adverse effects on traditional downstream beneficiaries (Křeček and Haigh 2019).

Mountain catchments in Central Europe are mostly forested. Willis (2002) categorised the main non-market costs and benefits of forestry practices with the priority in controlling runoff genesis: water yield (quantity and quality), runoff timing, flood-related risks, hydro-power generation, wildlife and recreational uses. To identify the above benefits, Foy et al. (2015) warned of serious uncertainties in present estimates of hydrological characteristics, especially in mountain regions. Environmental problems in mountain catchments are typically complex, uncertain, multi scale, and affect multiple actors and agencies. Therefore, the formulation of environmental services there requires flexible and transparent decision making approach that embraces a diversity of knowledge and values (Reed 2008). Ferrier and Jenkins (2009) point out different visions of what catchment management actually represents based on different sectoral perspectives. A key for the future is to

J. Křeček (✉) · E. Pažourková
Department of Hydrology, Czech Technical University in Prague, Prague, Czech Republic
e-mail: josef.krecek@fsv.cvut.cz; eva.pazourkova@fsv.cvut.cz

E. Pénzesné Kónya et al. (eds.), *Environmental Sustainability Education for a Changing World*, https://doi.org/10.1007/978-3-030-66384-1_6

look for synergies in these perspectives and to learning from the past's positive and negative experiences.

Citizen science has been effectively used in the field of catchment hydrology for many years and its potential depends on the adequate training and education of the involved lay people (Hand 2010; Pocock et al. 2018). Such field training should respect the basic principles formulated long ago by Comenius (1967) and include education in both knowledge and ethics. The aim of this paper is to evaluate the education strategy used by the citizen science project "Mountain Waters of Bohemia" (Earthwatch Institute 2012) upon integrated ecological monitoring and stakeholder dialogue for the restoration of headwater catchments and lakes in the Jizera Mountains (Northern Bohemia, Czech Republic), which have become degraded due to acid atmospheric deposition and by commercial forestry.

2 Material and Methods

2.1 Study Site

The Jizera Mountains Region is among the most important nature resource units in the country. This project was performed in headwater catchments at the upper plain of the Jizera Mountains (Northern Bohemia, Czech Republic, 50°40′–50°52′ N, 15°08′–15°24′ E, Fig. 6.1).

After World War II, consequences of rising soft coal combustion (air pollution, acid atmospheric deposition, dieback of spruce plantation) and intensive forestry practices (extended clear-cut, skidding timber by wheeled tractors) led to the degradation of headwater catchments.

In the 1980s, this area was strongly affected by the acid atmospheric deposition and die-back of spruce plantations (*Picea abies*); after the clear-cut of damaged spruce stands, grass dominated community *Junco effusi-Calamagrostietum villosae* spread there (Křeček and Hořická 2006). The upper plain of the mountains (approx.

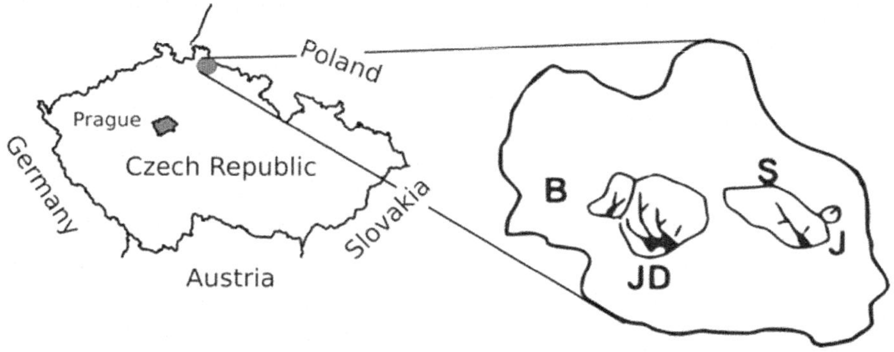

Fig. 6.1 The Jizera Mountains (Czech Republic)

200 km^2) is located above the elevation of 800 m (humid temperate climate, subarctic region, Dfc of the Köppen climate zones), where the mean annual precipitation ranges from 1290 to 1400 mm and the mean annual temperature from 4 °C to 5 °C. In a normal year, snow cover lasts from the beginning of November to the end of April, reaching the maximum snowpack of 120 cm (Tolasz et al. 2007). Surface flow dominates runoff genesis, groundwater bodies occur only in the shallow subsurface layers and the bedrock cracks. Because of the granite bedrock and poor shallow soils, local environments are extremely sensitive to acidification.

2.2 *Citizen Science*

In 1991–2012, voluntary citizen scientists participated in ground observations during field surveys (four to five teams a year of two-week duration, and, four to eight lay participants per team). These field expeditions were organised by the Earthwatch environmental programme (Earthwatch Institute 2012) using the concept of invitational research and education (Haigh et al. 2017). This approach was applied also by training of potential decision makers (corporate teams) involving in the "National debate on the management of mountain watersheds in the Jizera Mountains" (Chalupa and Křeček 1995).

After the standard preliminary selection done by the Earthwatch Institute, volunteers were instructed and trained in the field; and the accuracy of their results was assessed on a daily basis. Besides the catchment inventory, volunteers were used to gather information on stream water characteristics (*in situ* temperature, pH, and conductivity) and water sampling during some snowmelt and rainstorm events. Selected data from the volunteers were controlled under laboratory conditions. Simultaneously, this project focused also on the environmental education of the volunteers: after 2 weeks of field activities their knowledge and skills were tested, and the method of interviewing the research placement (Haigh et al. 2017) was used to assess their involvement.

2.3 *Environmental Services of Mountain Forests*

Multi resource forestry requires maintaining the forest as the biophysical system with the science-based dialogue between main stakeholders involved (Stollkleemann and Welp 2006). From the current status of knowledge, based on the Earthwatch citizen science expeditions, multi-criteria decision analysis (MCDA, Triantaphyllou 2000) suggested several basic steps toward the collaborative watershed management in two catchments of drinking water reservoirs Josefův Důl and Souš (Fig. 6.2).

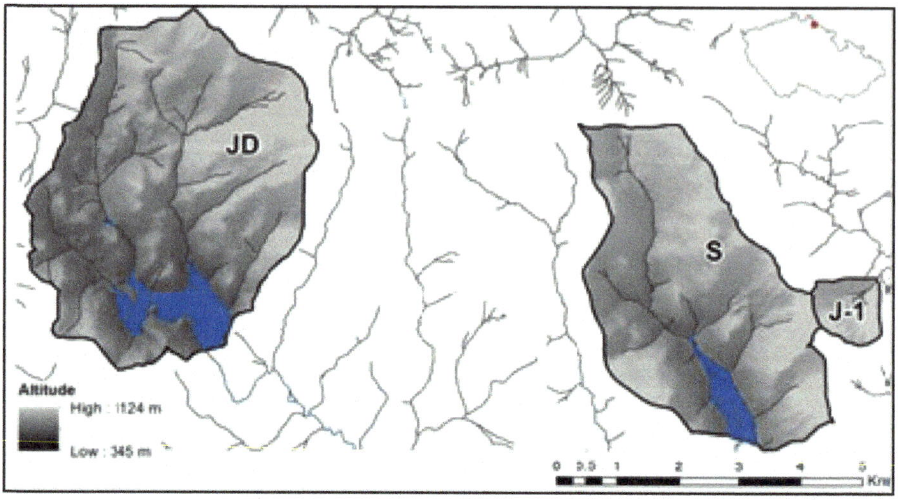

Fig. 6.2 Drinking water supply catchments Josefův Důl (JD) and Souš (S) and the Jizerka experimental catchment (J-1)

3 Results and Discussion

3.1 Integrated Ecological Monitoring

Citizen scientists played an important role in the extensive ground data gathering needed to address wider environmental consequences of the acid rain impact (Silvertown 2009). In the Jizerka catchment, the open field deposition of sulphur (S-SO_4) peaked in the late 1980s and showed a decreasing trend during the 1990s with the drop in atmospheric emissions of SO_2 (according to the 1985 Helsinki Sulphur Protocol) while the open field load of inorganic nitrogen (N–NO_3 and N-NH_4) did not change significantly (Křeček and Hořická 2006). Thus, during 1982–2018, the ratio between deposited nitrogen and sulphur has increased from 0.37 to 2.83. Ammonia presented 72 and nitrate 28% of the long-term load of inorganic nitrogen. Also, ground observations confirmed the linear hypsometric relation with precipitation sums, number of foggy days and fog drip amounts. The drop in catchment outflow of sulphur and nitrogen (Fig. 6.3) corresponds with the clear-cut of spruce stands (1984–1988) and reducing canopy area. Although the reforestation (mainly with spruce stands) in the Jizerka basin started immediately, in the year following harvest, the re-growth of forests was relatively slow. In consecutive years of 1992, 2002 and 2010, grass still dominated on 62, 37 and 19% of the catchment area (crown canopy closure of trees was below 0.3). This corresponds with the spreading of invasive herbaceous communities (namely *Calamagrostis* sp.) over the studied catchment with the defoliation of mature spruce stands (Křeček et al. 2019).

Similar correspondences between the drop in the acid atmospheric load and reducing forest canopy were identified in seasonal acidification (Fig. 6.4). The

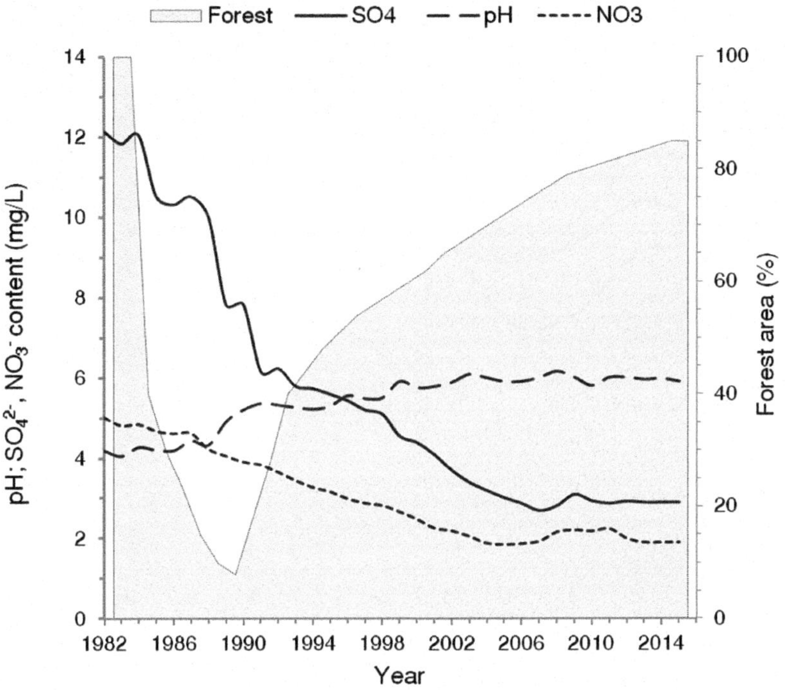

Fig. 6.3 Changes in the mean annual pH and contents of sulphate and nitrate in stream water with the regrowth of forests in the Jizerka basin (1982–2015)

recovery of stream water chemistry follows the drop in the acid atmospheric deposition in ca 5 years, but the revival of stream biota reflects these changes with a lag period of 10–15 years (Table 6.1). In 1994, the number of identified taxa of benthic organisms (36) still corresponds to a strongly acidified environment (pH below 4.2), while, in 2005, the number of taxa increased to 68 according to a moderately acidified environment (pH between 5.0 and 6.3).

3.2 *Citizen Science Efficiency*

Concerning the effectivity of using volunteers in the field research, Hodgson and Evans (1997) warned namely of reduced accuracy and increased response time by identifying characteristics of low ionic stream waters. The *in situ* data (water pH, conductivity) measured by volunteers showed a relatively good agreement with techniques under the laboratory conditions (testing 20 readings per expedition, the correlation coefficient r varied between 0.79 and 0.93 ($r_{crit} = 0.16$, $n = 100$, $p = 0.05$). By more complicated sampling benthic macroinvertebrates, volunteers assisted the professional project staff. In addition, volunteers allowed a greater detail in the

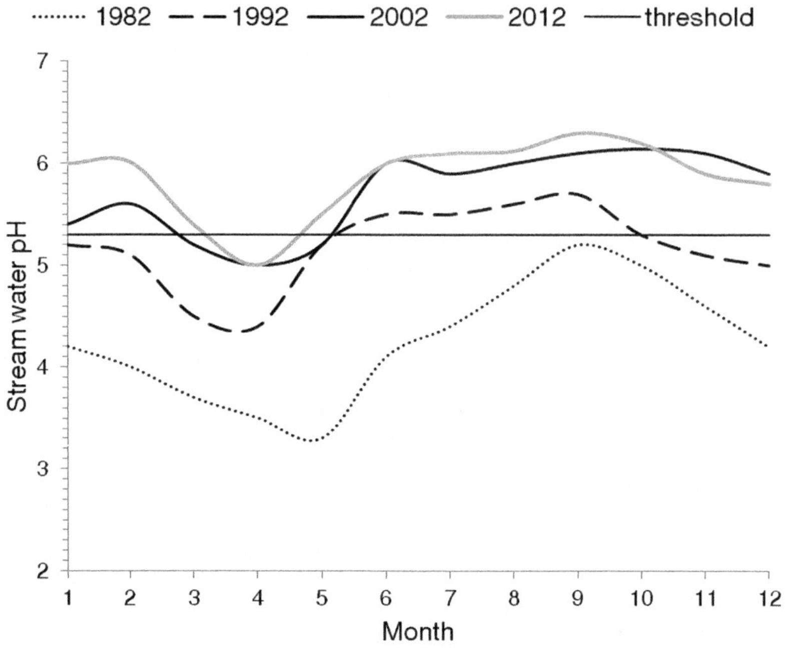

Fig. 6.4 Mean monthly pH of stream water at the outflow of the Jizerka catchment

Table 6.1 The number of identified taxa in benthic organisms at the Jizerka outlet: comparison between years 1994 and 2005

Taxa	1994	2005
Nematoda	–	1
Oligochaeta	2	4
Hydracarina	–	1
Crustacea	–	1
Ephemeroptera	–	3
Plecoptera	12	20
Megaloptera	1	–
Trichoptera	4	17
Diptera excl. Chironomidae	11	10
Chironomidae	4	5
Coleoptera	2	6
Total	36	68

sampling campaigns. The motivation of citizen participants played also an important role in the progress of their education evaluated by Earthwatch Institute (2012). The main motivation of volunteers is shown in Table 6.2. Insider interviews of volunteers participating in the field expeditions (cf. Haigh et al. 2017) were used to find the reflection in the training of volunteers and their involvement in research activities.

Table 6.2 Main motivation of citizen scientists to participate in field expeditions to the Jizera Mts. (1991–2012)

Motivation	Response rate (–)
Improving the environment	0.76
Extending the interdisciplinary knowledge	0.57
Socializing with other similar oriented people	0.42
Changing the office work for outdoor activities	0.38
Better understanding the local culture and history	0.21
Ecotourism	0.09

Table 6.3 Structured forestry zones proposed in the basins of Josefův Důl and Souš

	JD		S	
Structural forestry zones	(ha)	(%)	(ha)	(%)
Riparian buffer strips	140	8	50	4
Soil protection	40	2	60	5
Reducing evapotranspiration loss	1090	64	770	58
Fog drip control	360	21	430	33
Wetlands protection	70	5	0	0

3.3 Science-Based Stakeholder Dialogue

Core stakeholders represented the sectors of water management, forestry and nature conservation; additionally, municipalities, recreation business and local NGOs were involved in the roundtable debate (Chalupa and Křeček 1995).

According to the citizen science achievements, the multi-criterial decision process formulated the structured forestry in catchments of the water reservoirs Josefův Důl and Souš respecting the main target of water resources control there. Thus, five watershed zones were delineated with respect to their prevailing role in the runoff genesis: (1) riparian buffer zones, (2) zones of soil protection, (3) zones of evapotranspiration control, (4) zones of significant fog drip, and (5) wetlands (Table 6.3 and Fig. 6.5).

4 Conclusions

Citizen science seems to be an important mechanism to extend the integrated ecological monitoring in mountain watersheds. Environmental education is the ground for an effective involvement of citizen volunteers as well as for the science based dialogue of involved stakeholders to control the management of mountain forests.

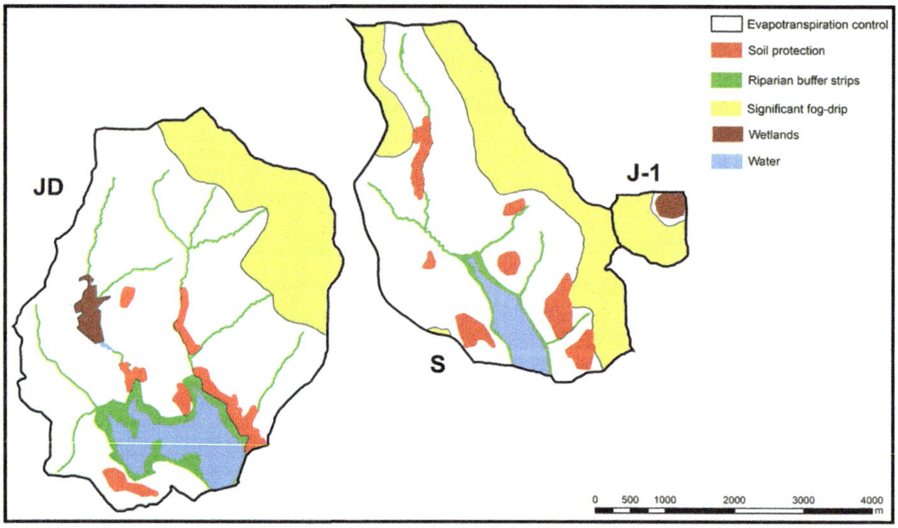

Fig. 6.5 Structured forestry zones in the basins of Josefův Důl and Souš

The forest percentage in catchments provides only limited information on their environmental services. 'Protective forests' controlling mainly the forest percentage (with some additional proscriptions of clear-cuts and drainage only) cannot ensure adequately the concept of environmental services. In a mountain catchment with priority of water resources recharge (quantity and quality), the structured forestry practices should respect five main priorities: protection of surface waters (buffer strips along reservoirs and streams), conservation of biodiversity (wetland-spots), soil protection (slopes over 30%), fog drip maintenance (elevations above 900 m), and evapotranspiration control. The precise delineation of these zones within a watershed depends on the morphological configuration.

To date, adoption of a mechanism for calculating the payment for environmental services is still rather distant and the methods proposed both complicated and inclusive of a relatively high risk of uncertainty (not least concerning climate change and nonpoint pollution). Meanwhile, the significant and continuing impacts of 'acid rain' on environmental services should be recognised and the acid atmospheric load of the catchment modified by controlling the canopy area and its roughness.

Acknowledgements This research was supported by the Ministry of Education (INTER-EXCELLENCE: INTER-COST LTC 17006, 2017–2020 and the Czech Technical University in Prague (SGS 18/120/OHK1/2T/11, 2018–2019).

References

Biswas, A.K., Tortajada, C. and Izquierdo, R. (2014). *Water Quality Management: Present Situations, Challenges and Future Perspectives*. Routledge, London (UK), 288 p.

Chalupa, M. and Křeček, J. (1995). National Debate on Management of Mountain Watersheds in the Jizera Mountains (Czech Republic). Forest-Water Foundation, Prague, 13 p.

Comenius, J.A. (1967). *The Great Didactic*. Russell and Russell, New York, 316 p.

Earthwatch Institute (2012). Earthwatch Annual Report. Earthwatch Europe, Oxford, 26 p.

FAO (2006). The New Generation of Watershed Management Programmes and Projects. Food and Agriculture Organization of the United Nations, FAO Forestry Paper 150, Rome, 128 p.

Foy, C., Arabi, M., Yen, H., Asce, A.M., Gironás, J. and Bailey, R.T. (2015). Multisite assessment of hydrologic processes in snow-dominated mountainous river basins in Colorado using a watershed model. *Journal of Hydrologic Engineering*, 20, 04015017, https://ascelibrary.org/doi/abs/10.1061/(ASCE)HE.1943-5584.0001130

Ferrier, R.C. and Jenkins, A. (2009). *Handbook of Catchment Management*. Wiley-Blackwell, Chichester, 560 p.

Haigh, M., Cotton, D. and Hall, T. (Eds) (2017). *Pedagogic Research in Geography Higher Education*. Routledge, Albingdon (UK), 335 p.

Hand, E. (2010). Citizen science: People power. *Nature*, 466, 685-687.

Hodgson, P. and Evans, J.G. (1997). Continuous pH, electrical conductivity and temperature measurement at Plynlimon: towards an integrated, reliable and low maintenance instrument system. *Hydrology and Earth System Sciences*, 1, 653–660, https://hal.archives-ouvertes.fr/hal-00304433/

Křeček, J. and Haigh, M. (2019). Land use policy in headwater catchments. *Land Use Policy*, 80, 410–414, https://doi.org/10.1016/j.landusepol.2018.03.043.

Křeček, J. and Hořická, Z. (2006). Forests, air pollution and water quality: Influencing health in the headwaters of Central Europe's "Black Triangle". *Unasylva*, 57, 46–49.

Křeček, J., Palán, L. and Stuchlík, E. (2019). Impacts of land use policy on the recovery of mountain catchments from acidification. *Land Use Policy*, 80, 439–448, https://doi.org/10.1016/j.landusepol.2017.10.018.

MEA (2005). *Ecosystems and Human Well-being: Synthesis. Millennium Ecosystem Assessment*. Island Press, Washington DC, 155 p.

Pocock, M.J.O., Chandler, M., Bonney, R. and Thornill, I. (2018). A vision for global biodiversity monitoring with citizen science. *Advances in Ecological Research*, 59, 169–223.

Reed, M.S. (2008). Stakeholder participation for environmental management: A literature review. *Biological Conservation*, 141, 2417–2431, https://doi.org/10.1016/j.biocon.2008.07.014

Silvertown, J. (2009). A new dawn for citizen science. *Trends in Ecology & Evolution*, 24(9), 467–471, https://doi.org/10.1016/j.tree.2009.03.017.

Stollkleemann, S. and Welp, M. (Eds) (2006). *Stakeholder Dialogues in Natural Resources Management, Theory and Practice*. Springer, Environmental Science, Berlin, 333 p.

Tolasz, R. et al. (2007). *Climate atlas of Czechi*a. Czech Hydrometeorological Institute, Prague, 256 p.

Triantaphyllou, E. (2000). *Multi-criteria Decision Making Methods: A Comparative Study*. Springer (Applied Optimization Series 44), Dordrecht, 290 p.

Willis, K.G. (2002). *Benefits and costs of forests to water supply and water quality*. Social and Environmental Benefits of Forestry, Report to Forestry Commission, Phase 2, Centre for Research in Environmental Appraisal and Management, University of Newcastle, p. 20.

Viviroli, D., Dürr, H.H., Messerli, B., Meybeck, M. and Weingartner, R. (2007). Mountains of the world, water towers for humanity: Typology, mapping, and global significance. *Water Resources Research*, 43, 1–13, https://doi.org/10.1029/2006WR005653.

Chapter 7
Education and Meals – Could Awareness-Raising Decrease the Food Footprint of Humankind?

Gabriella Rózsa, Attila Varga, and Andacs Noémi

1 Introduction

Today, agriculture is near to its global limits. Further increases in agricultural production could become impossible in the foreseeable future. Recent data suggests that the combined biomass of animals bigger than some kilograms (including humans) is circa 1.1 billion tons, which includes domestic animals (700 million tons, 63%), humans (300 million tons, 27%) and all remaining wildlife, including elephants and whales (100 million tons, 9%) (Harari 2017). Given the amount of Earth biomass that could now be converted to livestock or human bodies is limited, yet global human population is projected to increase 25% by 2050 (UN 2017); it is not difficult to imagine a scenario where humans and their animals simply replace all remaining wildlife and still face starvation. This is why exploring all possible avenues for reducing the need for agricultural products is vitally important both for human sustainability and the conservation of Earth's remaining biodiversity.

There are many feasible ways of reducing the human population's demand and land-take for agricultural products, such as consuming less meat and decreasing food intake to avoid obesity. However, among these possibilities, there is one that does not require changes either to the content or the quantity of what human's eat and so seems most easy to realise both culturally and psychologically. Simply, this is to reduce our food waste.

G. Rózsa (✉)
Eötvös Lóránd University, Budapest, Hungary

A. Varga
Hungarian Institute for Educational Research and Development, Eszterházy Károly University, Eger, Hungary

A. Noémi
Tudatos Vásárlók Egyesülete (Association of Conscious Consumers Hungary), Budapest, Hungary
e-mail: noemi@tve.hu

E. Pénzesné Kónya et al. (eds.), *Environmental Sustainability Education for a Changing World*, https://doi.org/10.1007/978-3-030-66384-1_7

Globally, the volume of food waste is huge and the role of households in food-supply chains is also enormous and creates serious health and environmental risks. Much of the problem is linked to the mind-set of our modern consumer society. Recent data from the Hungarian National Food Chain Safety Office show that the quantity of food waste per capita per year is nearly 70 kg, and almost half of this quantity is categorised as avoidable food waste, such as leftovers or food with expired 'best before' dates (NFCSO 2019). This means that food waste could be halved by encouraging less wasteful consumer behaviour.

Thus food production for humankind and their billions of pet animals is a major sector of the economy globally and has significant impacts on the biological resources on our planet (Williams et al. 2010). This is why human food habits could make a difference to our collective ecological footprint. This footprint is affected by two main factors: what we eat how and we prepare our food.

The first part of this chapter offers a short overview on how the theme of food is dealt with in the Hungarian public education system, where the main focus of food-education is health. It also describes current initiatives that aim to broaden the focus and integrate environmental issues in food education through projects involving school gardens, and community food education (e.g. Eathink).

The second part describes preliminary findings from a research on a project that aims to help communities rethink their food habits; this is Ökokör-Kamra (EcoTeam–Food). Ökokör-Kamra is a system of non-formal learning for adults supported in Hungary by the Tudatos Vásárlók Egyesülete (Association of Conscious Consumers). Its methodology is based on the work of the international, EU-funded, EcoTeams project coordinated by Global Action Plan International. This involves the creation of self-organising learning groups of 5–10 people, who learn together about food and environmental related issues. To date, 52 questionnaires and nine in-depth interviews have been collected from EcoTeam group coaches and participants. This chapter summarises the preliminary results of this research and focuses on which EcoTeam activities were most effective in changing participants' eating habits and, potentially, decreasing our global food footprint. It also identifies which activities require further development to be effective.

2 Food in National Core Curriculum of Hungary

An obvious way to address the challenge of food waste would be to introduce the topic into school curricula and so help the future generations to prepare to reduce their own food waste. In order to examine how this opportunity is used in Hungary we have analysed the text of the Hungarian National Core Curriculum (NAT 2012). Here, the expression of food and its synonyms are mentioned 50 times. The theme is present in all stages of school education but not in all subjects. The vast majority of the mentions are in science curricula (45). Five mentions are in the curricula of practical studies but there is no mention of food or any related subject in any social, human or art curricula. The main topics related to food are: nutrition – energy with about 50% of the mentions, health with ca. 30% of the mentions and food preparation with ca. 20% of the mentions. Sustainability issues are mentioned linked to food preparation but there is no detailed guidance provided for teachers about what has to be taught, especially about food waste reduction.

In summary, it could be stated the theme of food waste is present in the Hungarian National Core Curriculum, so the schools have to deal with it, but the theme is on the periphery even within the theme of food and this curriculum does not provide guidance for teachers about how to include the theme in their daily work. Social aspects of food consumption and food waste production are at best implicitly covered by this curriculum.

3 Food Education Projects in the Hungarian Educational System

The need for teaching guidelines and teaching materials on food waste was recognised by the National Food Chain Safety Office (NFCSO) which launched the project Wasteless in 2016 aiming to prevent food wastage in households (Wasteless 2019). Earlier, the programme had already targeted the adult population with a general information campaign, but, more recently, they have broadened their target group to include school pupils and children. The project publishes textbooks, exercise books, quizzes and other on-line games in order to help teachers deal with the theme of food waste.

Beside governmental actors, civil organisations also offer their help for schools about the theme of food waste. The "EAThink 2015 – eat local, think global" is a project of a consortium of non-governmental organisations from 12 European and two African countries, which has the objective of enhancing primary and secondary school students' and teachers' critical understanding and active engagement in global challenges, with a specific focus on food security and sovereignty, sustainable food systems and smallholder farming (EAThink 2015). Another example of a programme supported by non-governmental organisations is the Heroes of Responsible Dining Foundation, which promotes seasonal and local food in schools and kindergartens with special attention to both food waste and the waste production caused by food packaging (Worldchef 2020).

4 EcoTeams

Even if the school programmes about food waste were very widespread and effective their effect would be limited without the support of the adult population in the process of changing food preparation and eating habits. That's why, beside the general information campaigns mentioned earlier, there is a significant need for opportunities for informal learning where people could not just learn about the facts of food waste, but also could study new methods and approaches and could share their experience, talk about their own successes and failures and support each other in the process of change. These are the basic ideas behind the EcoTeam scheme.

The idea was created by Marilyn Mehlmann, who created the system on the bases of the experiences of her own small team (Mehlmann et al. 1996, 2010). The main idea was to begin the work of changing habits in favour for sustainability in small informal groups of 5–10 people. The conversations could start with the theme of reducing energy consumption and then, later on, deal with other complex issues of the problem. The goal is to consider the operation of households and to design behaviour with environmental protection in mind by examining and improving our everyday habits (Baur 2017).

The working principles of the eco-teams give an insight into their atmosphere and character:

1. No one can do everything, but we can all do something.
2. The work is shared by a group so that members can support and encourage one another and share experiences and ideas.
3. The approach is a step-by-step process that helps you to change your lifestyle by small adjustments that become good habits.
4. By measuring resource use, improvements can be monitored and encouraged during the course of the programme (Global Action Plan 2020).

The EcoTeam programme was introduced in many other countries including Hungary and was adapted to local conditions. At the start in Hungary in 2009, EcoTeams dealt with six topics (household economy, food, chemicals, energy, waste, water), which were extended with two new topics (health, transport) (e.g. Fahy et al. 2019). An EcoTeam course consisted of ten occasions: one opening and one closing session and eight thematic sessions one dedicated to each topic. Each EcoTeam was co-ordinated by a group leader. Group leaders were trained by Association of Conscious Consumers (TVE). One of the most important elements of the work of the EcoTeam group was the measurement of the decrease of food waste, energy water and chemical consumption, in participants' household during the course. The results from 2012 show that participants reported a 24% decrease in their food waste production on average (TVE 2012).

TVE organised a meeting for EcoTeam leaders in 2012 and gathered their opinions and suggestions. It turned out that, for most of the groups, the food theme took a lot more time than the other topics and there was a great need to deal with food more extensively. As a result, a new EcoTeam-Food programme was developed (TVE 2015) and launched in the spring of 2017 a food-focused training with fewer group meetings. The new EcoTeam-Food programme focused only on food-related issues alone. These were divided into three parts: food and health, food and waste, and food and climate change. A booklet with professional data was provided for the groups, but their emphasis was on trying, experiencing, measuring, observing, ideas and practices. The groups met five times weekly. The members' goal was to try to reduce the environmental impact of their meals long term by trying new practices and incorporating new routines. The group leaders demonstrated solutions that everyone can easily try and which illustrated the role we can play for reducing food waste.

5 Methods and Sample

Two methods were used to collect data about the effects of the EcoTeam or EcoTeam-Food programme on food waste reduction: semi-structured interviews and a questionnaire.

Interviews covered the following topics: background information, motivation, experiences and effect of participation in an EcoTeam group and feedback on EcoTeam method. Since the EcoTeam programme is not just about food issues, people often talked about other successes, experiences, and failures several times, and it could be difficult to keep the theme of the interview to food waste.

When choosing the interviewees, it was a condition for them to have, previously, participated in EcoTeam or EcoTeam-Food (although this could have happened at any time in the past years). TVE sent out a request in an e-mail for the former participants of EcoTeam and EcoTeam-Food groups. Finally, between November 2017 and January 2018, nine people took part in the interview, all of them attending EcoTeam training. Each interview took about an hour. Six were personal meetings usually in cafes or in one case in the home of the participant, three interviews were conducted via Skype. In each case, the participants decided about the circumstances of the interviews. There were seven female and two male interviewees. Seven of them lived in a family and two of them lived alone at the time of the interview.

Based on the experience of the interviews, a questionnaire was developed in order to gain data from a larger sample. The questionnaire was sent out to all of the former participants of the programme. Altogether 850 people were contacted, but only 52 (6%) people filled in the questionnaire. The questionnaire covered thematic units similar to the interview: background variables, effects of the EcoTeams,

Fig. 7.1 Age of the participants ($n = 52$) (%)

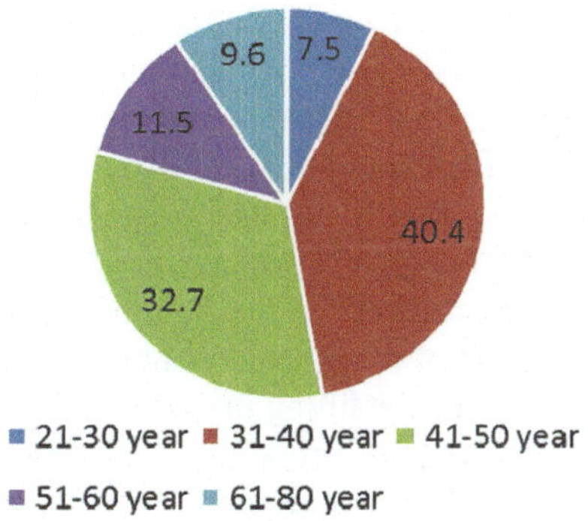

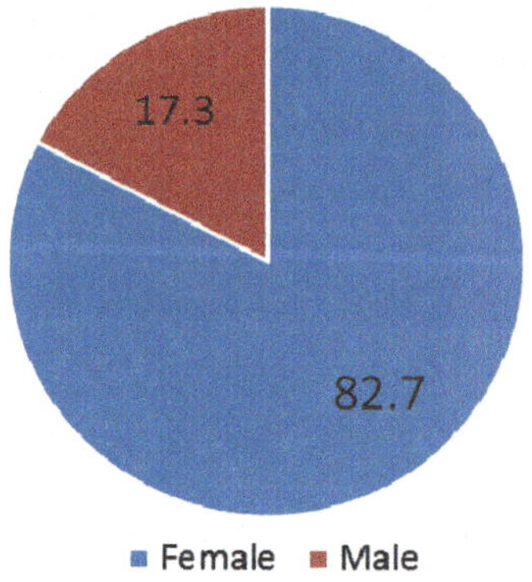

Fig. 7.2 Gender of the participants ($n = 52$) (%)

recommendation for improving Eco-Teams. Figures 7.1 and 7.2 show some basic features of the sample.

6 Results

The questionnaire used in our study was developed on the basis of the interviews, it covered almost the same topics, and interviewees filled in the questionnaire too. The results of our study is based, mainly, on the data gathered by the questionnaires.

A huge majority of the respondents (94.1%) gave positive feedback about EcoTeams and more than three quarter of them (78.9%) reported long-term changes in their habits related to food waste. The following quotes from the interviews demonstrate these changes:

> … EcoTeam has opened my eyes – I have not been an environment destroyer intentionally but my eyes have opened that packed food and unpacked food means a different approach
>
> Do we really need those things we 'needed' so far? We thought over it. I realised that while I used to think to myself that I am environmentally conscious, but not…

Encouraging the involvement of family members was successful: about two-thirds (63.46%) of respondents said they had received support from their families. Children and spouses were open, but of course there were resistance in some households, for example in the case of composters (to be in the house, or outside or to be collected at all). Almost all (90.2%) respondents told about their training to their friends and relatives. They shared concrete ideas, tips and cost-effective techniques that they learned. They also talked about the good mood and the community feeling of the Teams and encouraged others to join. A little bit more than one fifth of the respondents (22%) had already run their own EcoTeams.

According to the results, measurement of the changes in the environmental impact of the households turned out to be one of the great difficulties of the training because, the filling in of the questionnaire, was difficult for the participants and they think of it as 'prodding'. Only 56.9% of the respondents indicated that the proposed methods for measurement were feasible. More than half (54.2%) mentioned the lack of time and more than one third (35.4%) referred to the complexity of the questionnaire, as reason they were not able to complete the task. Obviously, a more simple and user friendly questionnaire must be developed.

Participants also reported some negative opinions from their social environment. They met with people who were absolutely not interested in the work of the EcoTeams and some who considered the whole thing a total waste of time.

At the end of the questionnaire and the interviews, the respondents were encouraged to make suggestions, and share their thoughts on their training and the project. Almost two thirds of the respondents commented. Most of the suggestions were about the dissemination of the project. The majority of the respondents suggested organising more teams and training more team leaders. There were also a few concrete suggestions about improving the work of the teams, which also could be relevant to the work of the teams but other awareness programmes too. So, it is important to overview these answers.

One advised on methods of measurement. She suggested a more detailed description and the use of alternative ways of measurement – for example, taking photos of the household waste would be easier for those who are not very interested in (or good at) numbers.

One parent emphasised that the changes in the family could cause many conflicts and suggested an alternative methodology (which will be tested in a future EcoTeam), which involves both of the adults of the household. The aim is to ensure that the principles and goals of the EcoTeam are recognized by both parents and, hence, communicated to the children in a more coherent way.

Another respondent highlighted the need of transforming traditional household roles, because most of the EcoTeams members were women. It was recommended that the role of men in households should be increased and that these kinds of eco-topics should be considered by the whole family, not only by women.

One respondent shared a negative, personal experience. On one occasion the respondent gave a presentation to the group, which caused some disagreement with the team leader. The leader scolded the respondent, which caused some bad feeling and the respondent missed the next two meetings. Positive steps made by the participants should be supported while better training of team leaders is crucial (e.g. Boniwell 2012).

7 Conclusions

Although this research is limited in many ways, it has demonstrated that it is possible to develop and realize educational programmes aiming to reduce food waste production and thus the ecological footprints of households. It also demonstrates the potential of creating and providing facilitation for learning about food waste in informal groups (EcoTeams) and establishes that this could be an effective way to reduce food waste.

On the other hand, experience has shown that there is a significant need for the improvement in the organisation of EcoTeams. The low number of questionnaire respondents could be an indicator of ineffective communication between the organisers and participants of the EcoTeams. It could also indicate that, among those who did not respond, there were more people who had negative experiences with eco-training. This is why special attention is needed for the continuous collection of feedback from the participants and for the improvement of the organisation and leadership of EcoTeams. The tasks should be very clearly communicated to participants, the methods of proposed measurement should be as simple as possible, and encouragement and positive feedback should be the default method of giving feedback to the participants. It is worthwhile to mention that TVE will launch the new generation of EcoTeams this year and that its special focus will be household chemicals. In part, thanks to our experiences, the new initiative will pay greater attention to developing as simple measurement package as possible. Finally, we hope that this project and our results will encourage other people both to organise their own EcoTeams and also to share their experiences and so improve the effectiveness of this strategy.

References

Baur, M. (2017). *Empowerment is the key to sustainable behaviour change* – Interview with Marilyn and Alexander Mehlmann. Scaling4good, 2019.02.08, https://scaling4good.com/2017/10/12/empowerment-sustainable-behaviour-change-interview-marilyn-mehlmann/.

Boniwell, I. (2012). *Positive Psychology in a Nutshell (3e)*. Open University Press (McGraw-Hill Education), Maidenhead, Berks, UK.

EAThink (2015). EAThink2015 – eat local, think global, https://eathink2015.org/en/

Fahy, F., Heaslip, E., Hajdinjak, M., Asenova, D., Bouuaert, M.C., Tkalec, T., Živčič, L., Bellmallem, R., Čoklc, K., Gomes, C. and Slezak, G.H. (2019). Country Report: Hungary. ENERGISE (European Network for Research, Good Practice and Innovation for Sustainable Energy: 727642). EU Commission (Horizon 2020), Brussels.

Global Action Plan (2020). EcoTeams: https://www.nus.org.uk/PageFiles/8910/GAP%20Eco-team%20generic%20information.pdf

Harari, Y. (2017). *Ecology*. Letöltés dátuma: 2019.02.07, https://www.ynharari.com/topic/ecology/.

Mehlmann, M., Bingel, E. and Thunberg, B. (1996). *EkoTeam – Ta makten över miljöutvecklingen!* (EkoTeam, a Workbook for Individuals and Households Wishing to Live More Sustainably). Stockholm, GAP Sverige.

Mehlmann, M., McLaren, N. and Pometun, O. (2010). Learning to live sustainably. *Global Environmental Research*, 15, 177–186.

NAT (2012). 110/2012. (VI. 4.) *Korm. rendelet a Nemzeti alaptanterv kiadásáról, bevezetéséről és alkalmazásáról.* (Governmental decree on the publishing and introduction of National Core Curriculum) http://ofi.hu/sites/default/files/attachments/mk_nat_20121.pdf.

NFCSO (2019). *Maradék Nélkül* (Wasteless), http://maradeknelkul.hu/.

TVE (2012). *Zöldítés örömmel 12-13* (Greening with joy). Tudatos Vásárlók Egyesülete – Association of Conscious Consumers, https://issuu.com/tudatosvasarlok/docs/full_ujcimlap.

TVE (2015). *Ökokör Kamra kézikönyv csoportvezetőknek* (Handbook for EcoTeam-food team leaders). Tudatos Vásrlók Egyesülete – Association of Conscious Consumers: 2019.02.08,

https://tudatosvasarlo.hu/sites/tudatosvasarlo.hu/files/okokor_kamra_-_kezikonyv_csoport-vezetoknek_0.pdf.
UN (2017). World population projected to reach 9.8 billion in 2050, and 11.2 billion in 2100. United Nations, https://www.un.org/development/desa/en/news/population/world-population-prospects-2017.html.
Wasteless (2019). School Programme. https://maradeknelkul.hu/en/school-programme/
Williams, W.J., Hughes, J.K., Black, M. and Murphy, R. (2010). Energy and the food system. *Phil. Trans. R. Soc.* B, 365, 2991–3006, https://doi.org/10.1098/rstb.2010.0172.
Worldchef (2020). Heroes of Responsible Dining Foundation. https://www.worldchefs.org/about-us/our-partners/fc686be587a7a28df4b881623b248535cfa03b37.

Chapter 8
Public Educational Initiatives for Promotion of Knowledge About Rare Plants in Ukraine

Mykyta Peregrym, Oleksij Vasyliuk, and Olesya Bezsmertna

1 Introduction

The natural flora of Ukraine is one of richest in Europe. It counts about 5000 species of vascular plants (Mosyakin and Fedoronchuk 1999). The existence of such diversity is possible due to natural conditions of three botanical-geographical zones (Forest, Forest-Steppe and Steppe Zones) and two mountain zones (Crimean Mountains and Carpathians), which have been formed in post-glacial time (Mosyakin et al. 2005). However, anthropogenic influence on natural ecosystems has been growing very fast during the last 300 years, as a result their areas have reduced. For example, steppes occupied about 40% of the modern territory of Ukraine, but, unfortunately, presently only about 3% of the natural and semi natural steppes of the country remain, because of ploughing, mining operations, excavations etc. (Korotchenko and Peregrym 2012). Another example shows the dynamics and quantity of rare and endangered species in flora of Ukraine. The second edition of the Red Data Book of Ukraine lists 439 species of vascular plants (Shelyag-Sosonko 1996), but there are 611 species of vascular plants listed in the third edition (Didukh 2009). Of course, it is important to note that the conceptions of these editions have differences but there is no doubt that situation with plant diversity has

M. Peregrym (✉)
Eszterházy Károly University, Eger, Hungary

NGO "Ukrainian Nature Conservation Group", Kyiv, Ukraine
e-mail: Mykyta.Peregrym@uni-eszterhazy.hu; http://uncg.org.ua

O. Vasyliuk
NGO "Ukrainian Nature Conservation Group", Kyiv, Ukraine
http://uncg.org.ua

O. Bezsmertna
Department of Botany (and Oleksii Vasyliuk's Lab), Taras Shevchenko National University of Kyiv, Kyiv, Ukraine

E. Pénzesné Kónya et al. (eds.), *Environmental Sustainability Education for a Changing World*, https://doi.org/10.1007/978-3-030-66384-1_8

become worse (Didukh 2010). Besides, if the reasons for the rarity of these species (which are specified in the last edition of the Red Data Book (Didukh 2009) are analyzed, it can be seen that most plants have become rare or endangered species due to the direct impact of anthropogenic factors on their habitats or their populations. Only a very small number of species are threatened by climate change, desertification or sea level rise.

At the same time, presently, the majority of Ukrainian citizens are not able to name even 10–15 of the species that are under protection in the country and listed in the Red Data Book of Ukraine. However, many people like spending their free time near a river, in a forest or along seashore, where they often meet protected species and the risk is that they will unconsciously collect them or destroy their habitats. This happens more than it should because information about rare plants is not widely available, while the Red Data Book exists only as a limited number of expensive hard copies. The Ministry of Ecology and Natural Resources has created a thematic website (http://redbook-ua.org), but this only partly resolves the problem because of information inaccessibility; many countryside schools have no Internet access or a restricted number of PCs. Therefore, the promotion of knowledge about rare plants, their value and importance of their conservation in Ukraine is a real problem, the solution of which may improve the efficiency of plant conservation as well as decreasing anthropogenic pressure on natural ecosystems in the country.

2 History of Plant Red Listing in the Country

The history of plant conservation in the modern territory of Ukraine extends back more than 130 years. It began, in 1886, with the foundation of the first protected area near Peniaky village (now Brody district, Lviv region), when count Włodzimierz Dzieduszycki decided to preserve 22.4 ha of a beech forest for scientific and aesthetic purposes (Melnik 2000). A little later, in 1889, Friedrich von Falz-Fein began to preserve a plot of virgin steppe in the southern of the modern territory of Ukraine, Today, this date is accepted as the foundation time of the Biosphere Reserve "Askania-Nova" (Zalesskyi 1915).

However, little attention was paid to the rarity of individual species and the importance of their protection until 1927, when a list of rare plants of Taganrog and Mariupol districts (between the Mius and the Kalmius rivers) was completed (Lavrenko 1927). The first list of rare, endemic and endangered plants for the whole country was prepared by M.I. Kotov (Kotov 1962) and quickly updated and improved by V.I. Chopyk (Chopyk 1963).

The initiation of Red Listing for plants began in the USSR with the first edition of the IUCN Red List in 1964, and the Soviet authorities began to take into account the importance of the active conservation of some species. As a result, the several official and scientific editions of the Red Data Book of the USSR were published (Borodin et al. 1978, 1984; Takhtajan 1975). In those times, the Ukrainian SSR was a regional leader in Plant Red Listing because V.I. Chopyk prepared two editions of the list of rare plants of the republic (Chopyk 1970, 1978), which became the basis for the Red Data Book of the Ukrainian SSR (Sytnik 1980).

After independence, two new editions of the Red Data Book for the Ukraine were published (Didukh 2009; Shelyag-Sosonko 1996) and regional lists of rare plants were approved for the majority of administrative regions (Andriyenko and Peregrym 2012). Thus, Ukraine has a rich experience in Plant Red Listing, which went beyond just scientific work from 2002, when the law about the Red Data Book of Ukraine was accepted. Nowadays, this law is one of the main and most effective tools for nature conservation because, by this law, the presence of a Red Data Book listed species in borders of any natural or semi natural territory constitutes a basis for the creation of a protected area, where destruction or damage to individuals of a rare species or destruction affecting its natural habitat become punishable offences. According to the law, information about rare plants of the country must be updated every 10 years and be published in new editions of the Red Data Book.

Hence, today Ukraine has a strong scientific and legal foundation for effective plant conservation as well as for the formation of a national network of protected areas, which is continuing to develop. However, the weak link is the limited level of knowledge in local populations, which persists because of the limited availability of suitable popular science literature with information about rare plants.

3 Public Ecological Actions for Plant Conservation During After Independence

Since the proclamation of Ukraine's independence, one of the main interests of Ukrainian nature conservation NGOs has been plant conservation. Given the large diversity of the Ukrainian flora, most species of which can be identified only by professional botanists, public nature conservation activities have been confined to a few of the most recognizable 'branded' or 'flagship' species, which have been used to stimulate popular concern about rare plants. For example, such plants in Ukraine include species from genera *Crocus* L., *Galanthus* L., *Leucojum* L., *Pulsatilla* Mill., and others that are picked for bouquets or dug up for transplantation into private gardens or use for food.

Public activity for the study and conservation of rare plants has taken many different forms across the country: art competitions, field excursions, thematic lessons in schools. Some NGOs have published popular books, calendars, posters etc. Sometimes, such initiatives have been supported by the Ministry of Education and Science of Ukraine through state institutions like the National Ecological and Naturalistic Center of Youth. For example, an annual contest, "Pervotsvity" (the name means the first spring flowers), has been organised among children's clubs and organisations. In fact, this has generated a lot of valuable information about the distribution of rare plants in Ukraine and some of the competition entries have led to the creation of new protected areas.

The most interesting historical precedents have been the formation of the Nature Protection Brigades, which have united students interested in nature conservation. The movement originated during the last years of the Soviet Union, but its popularity increased after 1991. Up to 45 such Brigades have existed in Ukraine at different times (Vasylyuk and Ilnytska 2010). One of their traditional activities has been to struggle against the illegal sale of early spring flowers. Students, with or sometimes without cooperation from the police, have stopped cases of such trade in many cities. Moreover, their work has exposed more than 10 illegal plant smuggling routes across and beyond Ukraine. In addition, these Student Nature Protection Brigades have written hundreds of articles on this theme for local media. This approach has proved very effective; as a result, this illegal trade was stopped, completely in Kyiv, and almost completely in the next four biggest cities in Ukraine. This activity has also created a tradition – most Ukrainian mass media consider it their duty to publish at least one article per year about the inadmissibility of the trade of early spring flowers. Another result has been to add many species of early spring blooming plants, which had not been listed in the Red Data Book of Ukraine, to regional red lists of rare plants.

It is important to mention that members of the Kyiv Nature Protection Brigade were involved in reintroduction and repatriation researches on plants which they had confiscated during 2002–2012. Firstly, students organised field expeditions to natural habitats of early spring blooming species for studying their particularities. Next, they looked for similar places for planting of the confiscated bulbs or plants. In some cases, these plants were planted in Kyiv botanical gardens, so creating artificial reserve populations. The most interesting example was *Cyclamen kuznetzovii* Kotov et Czernowa, a local endemic known from just one location in the Crimean Mountains (Shelyag-Sosonko 1996), which nevertheless, despite its rarity, regularly became an object of illegal trade. Students of the Kyiv Nature Protection Brigade studied the specific conditions of the *C. kuznetzovii* natural habitat, located areas with similar conditions elsewhere in Crimea, and there established new populations (Parnikoza and Inozemtseva 2009; Shevshenko et al. 2008). As a consequence, the chances of preserving the species in nature have increased significantly.

Unfortunately, the activities of these youth and student nature protection unions have decreased in recent years. Modern multimedia technologies and interactive communication may be required to renew this interest. Nevertheless, new forms of study and protection of plants arising from the involvement of public have already appeared in Ukraine, especially those using social networks. Detailed information are given below.

4 Public Educational Initiatives: Popular Science Books, Social Networks and Two Open-Access Data Banks

4.1 Edition "50 Rare Plants"

Taking into account the problems mentioned above, a series of popular science publications about rare plants for every region of Ukraine has been initiated to promote greater knowledge about these unique species. This has been named "50 rare plants", because its authors chose 50 rare species, mostly listed in the Red Data Book of Ukraine, for every region and given short morphological descriptions, habitat characteristics, data about their locations within protected areas, and the main threats to the future existence of their populations. An example of a bottom part of a page with colour pictures and distribution maps is shown in Fig. 8.1.

The regional approach is clearly appropriate for Ukraine, which has a total area of 603,700 km^2, or 5.7% of Europe's land area, and which includes three natural zones and two mountain systems. In addition, azonal habitats are widely spread across Ukraine. Thus, the flora is quite distinct in different administrative regions.

The first book "50 rare plants of Luhansk region" was published in 2014 by public activists and professional botanists with grant support from the United Nations Development Programme (Peregrym et al. 2014). The second book "50 rare plants of Donetsk region" was written together with scientists from the National

Fig. 8.1 A typical page from the series of editions "50 rare plants". This example of *Hyacinthella pallasiana* (Steven) Losinsk comes from "50 rare plants of Luhansk region" (Peregrym et al. 2014)

Natural Park "Meotida" in 2017 (Peregrym et al. 2017), and supported by public fundraising. The third publication "50 rare plants of Cherkasy region" was prepared in 2018 with participation of botanists from the Cherkasy Bohdan Khmelnytsky National University, Kaniv Nature Reserve and M.G. Kholodny Institute of Botany of the National Academy of Sciences of Ukraine (Vasyliuk et al. 2018). This publication became a reality thanks to financial support from the Cherkasy region.

This series of publications has become popular among school pupils, naturalists, and professional botanists thanks both to its concept as well as its approach to preparing the distribution maps of plants. It was preceded by long-term work on the creation of a database, which has involved collecting data from herbariums, scientific literature and personal observations of local experts and naturalists for every edition. Collaboration with local experts and naturalists has been a big success, because it allows collecting a lot of new data, and even the discovery of rare species and new locations that were not known in some regions, as well as having the results of verifications at well-known locations. As an example, Fig. 8.2 presents a graphic that compares the map of *Calophaca wolgarica* (L. f.) DC. distribution from the last edition of the Red Data Book of Ukraine (Didukh 2009) with the version updated for "50 rare plants of Donetsk region" (Peregrym et al. 2017).

A 1000 hard copies were printed for each of these regions and distributed to local libraries, schools and universities. Electronic copies were made available online. In collaboration with botanical experts, further works are in preparation for Odesa, Sumy and Kyiv regions. Hence, this public educational initiative for promotion of knowledge about rare plants in regions of Ukraine has become an important initiative in the campaign for plant conservation.

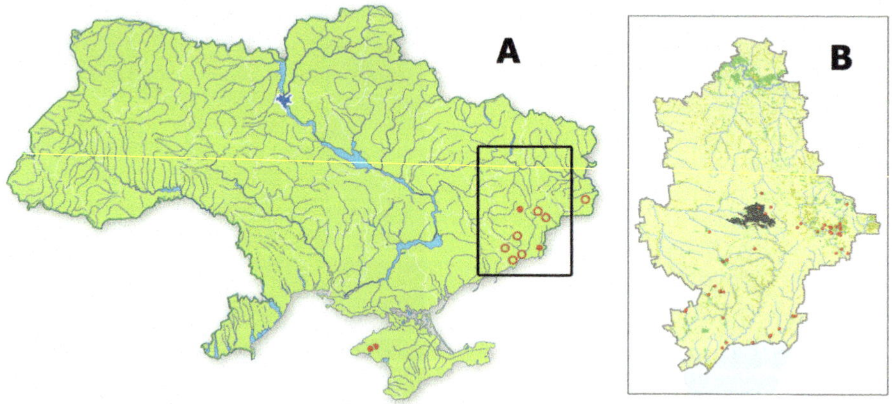

Fig. 8.2 Comparison of cartographic materials from the Red Data Book of Ukraine (Didukh 2009) and the edition "50 rare plants of Donetsk region" (Peregrym et al. 2017); example of distribution of *Calophaca wolgarica* (L. f.) DC

4.2 Promotion of Knowledge About Rare Plants Through Social Networks

There is no doubt that the initiative "50 Rare Plants" makes a great contribution to increasing the level of ecological knowledge of local populations, but it is not enough to significantly change public understanding of importance of plant conservation to Ukrainian society. Open sources where people can learn the latest information about the flora of Ukraine, plant conservation activity and different botanical and ecological events, have been absent in the country for a long time. However, this situation was changed on January 4, 2016, when the Facebook group "Flora of Ukraine" (https://www.facebook.com/groups/floraofukraine/) was created through a public initiative of several botanists.

The main aim of this Facebook social group is to collect a picture database of all native, alien and cultivated plants of Ukraine and so introduce people to the huge plant biodiversity of the country. This also shares information about rare plants, especially from the Red Data Book of Ukraine, protected areas, new publications, field expeditions, scientific conferences and workshops, special dates etc. Today, the Flora of Ukraine community has become a platform for communication between professional botanists, ecologists and amateurs. On its pages, there are discussions on important environmental issues that are interesting not only to professionals, but also for many ordinary citizens who want to learn more about what and how professional botanists serve the modern world. A slightly veiled goal of the project is to show that botanical science is interesting, creative, and important to the modern world.

The activity of the group observes certain rules. The most important of them is the publication of only album and one new species each day. The album has to contain several photos (typically from 2 to 10): the general view of the plant, macro photos with taxonomic features, and species habitats. Each photo has to be captioned with the name of the plant species, location and time when the photos were taken. Thus, having searched for a selection of photos in a certain species, group members can find out the conditions of the species growing, how it looks like and where it is distributed. One of the moderators is responsible for this work during the week and it should be noted all the photos are copyrighted because they were made by moderators during various expeditions and other scientific trips.

The team of moderators contains about 26 experts-enthusiasts from different regions of Ukraine: Kyiv, Cherkasy, Chernigiv, Chernivtsi, Dnipro, Ivano-Frankivsk, Kharkiv, Kherson, Khmelnytsky, Lviv, Luhansk, Lutsk, Odesa, Pervomaysk, Rakhiv, Sumy, Uman, Vinnytsia, and Zaporizhzhya. But the list is gradually expanding. The overwhelming majority of moderators are interested and highly qualified professional botanists from national and regional scientific, educational and environmental universities, institutions and research centres. Already, albums have been published for more than 1700 species which is about 35% of vascular plant flora of Ukraine (https://www.facebook.com/groups/floraofukraine/photos/?filter=albums). At present, the group has almost 3500 subscribers and is constantly growing. With its ex-pansion the Facebook group "Flora of Ukraine", its informative content and its value grows too. Followers often leave comments with their own photos and questions like "Have I seen the same plant?", "This plant is also growing at...", etc. Such additional information from ordinary interested citizens is particularly valuable because it can extend the knowledge base on species distribution, especially rare and endangered ones.

For Ukraine this is an unprecedented case of popularization of both knowledge about plants and science as a whole, especially in view of the fact that the modern illustrated reference books and atlases of plants are absent from the country. Currently, two further similar groups exist in the Ukrainian segment of Facebook: "Ukrainian Botanical Group" (https://www.facebook.com/groups/flora.ukraine/) and "Plant World of Ukraine" (https://www.facebook.com/groups/roslynnyy.svit.ukrayiny/). However, their content is different from the Facebook group "Flora of Ukraine" and often directed to helping the identification of plants using pictures posted by their visitors. Meanwhile, the "Ukrainian Botanical Group" specialises in sharing information about results of expeditions and trips of Ukrainian professional botanists and amateurs abroad.

Thus, a good platform for collecting data about the plant diversity of Ukraine and for the promotion of knowledge about it has been created by public activists on social networks. Received information is already used for preparing scientific and popular scientific publications, especially about rare and endangered plants. In addition, the site has become an important stage for the early development of citizen science in Ukraine and a foundation for digital databases of biodiversity.

4.3 Botanical Resources for Ukrainian Citizen Science

There have been two biodiversity databases in Ukraine since 2017: "UkrBIN: Ukrainian Biodiversity Information Network" (http://www.ukrbin.com/) and "Biodiversity of Ukraine" (http://dc.smnh.org/biodiversity-data-centre.html). The first was created as a public project in collaboration with two Ukrainian academic institutions: the I.I.Schmalhausen Institute of Zoology, National Academy of Science of Ukraine (Kyiv) and the Institute of Ecology of the Carpathians, National Academy of Science of Ukraine (Lviv). The second was created by the State Museum of Natural History, National Academy of Sciences, Ukraine (Lviv), within the framework of the applied-scientific project "Creation of the museum-information resource as a base of the regional action plans on biodiversity protection".

Currently, the database "Biodiversity of Ukraine" is developing slowly, and aimed at professional experts, others must access data using a special form on the website. The UkrBIN site is absolutely different. Here the database fills out through its website, where everyone can post results of observations using its Facebook page (https://www.facebook.com/groups/ukrbin.info/), as well by copying data from other open access sources such as the Facebook groups "Flora of Ukraine", "Ukrainian Botanical Group" or "Plant World of Ukraine" etc. UkrBIN's goal is stated as maximizing the efficiency and availability of as much data as possible for the benefit of both amateur and professional biologists. It is successful; in August 2020 alone, participants reported more than 422 thousand observations in Ukraine. The observations of each participant are cumulated with those of others in the UkrBIN international network. UkrBIN then shares these observations with the global community of lecturers, land surveyors, environmentalists and nature conservationists. These data form an important reservoir of knowledge about the distribution of biota throughout Ukraine. Thus, a small public initiative for the promotion of knowledge about rare plants has become a national level resources of biodiversity data and a source for the development of citizen science.

5 Conclusions

The awareness and knowledge of Ukrainian citizens about rare and endangered plants remains low at the present. Few know much about those special species that are under protection in the country and listed in the Red Data Book of Ukraine. Although the law about the Red Data Book remains the main and most effective tool for nature conservation, the presence of a Red Data Book species within the borders of any natural or semi natural territory can provide for the creation of a protected area and public educational initiatives aimed at promotion of knowledge about rare plants in Ukraine have been initiated. These include: the publication of a series of popular science books, "50 Rare Plants", about the rare and unique species of, eventually, every region of Ukraine and the promotion of knowledge about local, especially rare and endangered, plants through special informational groups in social networks. A side effect of the latter has been building the involvement of amateur citizen scientists in local projects engaged in data collection about biodiversity in Ukraine. The latest feedback shows that these public initiatives are regarded as successful by specialists and strongly contribute to raising the public awareness of rare and endangered plant species. Unfortunately, the current military, political and economic crises in Ukraine have led to funding cuts for its research and education centres, a loss of government attention to importance of nature conservation, as well as a brain drain of scientists to other countries. However, these new initiatives grant information and opportunities to everyone interested in plant diversity to make their own scientific contributions.

Acknowledgement Authors are grateful to Grigoriy Kolomytsev (I.I. Schmalhausen Institute of Zoology of the National Academy of Sciences of Ukraine, Kyiv, Ukraine) and Dariia Shyriaieva (M.G. Kholodny Institute of Botany of the National Academy of Sciences, Kyiv, Ukraine) for their advice for improving of our manuscript as well as for their collaboration in preparation of publications within our project "50 rare plants".

References

Andriyenko, T.L. and Peregrym, M.M. (Eds) (2012). *Official lists of regional rare plants of administrative territories of Ukraine (reference book)* (In Ukrainian). Alterpress, Kyiv.

Borodin, A.M., Bannikov, A.G., Sokolov, V.Y. et al. (Eds) (1984). *The Red Data Book of the USSR. Rare and endangered animal and plant species* (In Russian). Lesnaya promyshlennost', Moskow.

Borodin, A.M., Bannikov, A.G., Syroechkovskiy, Y.Y. et al. (Eds) (1978). *The Red Data Book of the USSR. Rare and endangered animal and plant species* (In Russian). Lesnaya promyshlennost', Moskow.

Chopyk, V.I. (1963). *Rare Plants of the Ukrainian SSR and Their Conservation* (In Ukrainian). Naukova Dumka, Kyiv.

Chopyk, V.I. (1970). *Rare Plants of Ukraine* (In Ukrainian). Naukova Dumka, Kyiv.

Chopyk, V.I. (1978). *Rare and Endangered Plants of Ukraine* (In Ukrainian). Naukova Dumka, Kiev.

Didukh, Y.P. (Ed.) (2009). *The Red Data Book of Ukraine. Plant Kingdom* (In Ukrainian). Globalconsulting Press, Kyiv.

Didukh, Y.P. (2010). The Red Data Book of Ukraine. Plant Kingdom. An afterword (In Ukrainian). *Ukrainian Botanical Journal*, 67(4), 481–503, http://nbuv.gov.ua/UJRN/UBJ_2010_67_4_3.

Korotchenko, I. and Peregrym, M. (2012). Ukrainian Steppes in the Past, at Present and in the Future. *In:* Werger, M.J.A. and van Staalduinen, M.A. (eds), *Eurasian Steppes. Ecological Problems and Livelihoods in a Changing World.* Springer, Dordrecht, pp. 173–196. DOI:https://doi.org/10.1007/978-94-007-3886-7_5

Kotov, M.I. (1962). Rare, endemic and endangered species of plants of the Ukrainian SSR and the need of their protection (In Russian). *Nature Protection and Conservation in the* USSR, 7, 50–53.

Lavrenko, Y.M. (1927). A list of rare plants which have been found within Taganrog and Mariupol districts between the Mius and the Kalmius Rivers (In Ukrainian). *The News of the Kharkiv Agricultural Institute*, 8/9, 41–48.

Melnik, V.I. (2000). *Rare species of the flora of plain forest of Ukraine* (In Russian). Phytosociocenter, Kiev.

Mosyakin, S.L., Bezusko, L.G. and Mosyakin, A.S. (2005). Relicts, refugia, and migration routes of plants in Europe in the Pleistocene-Holocene: A brief overview of phylogeographic evidence (In Ukrainian). *Ukrainian Botanical Journal*, 62(6), 777–789.

Mosyakin, S.L. and Fedoronchuk, M.M. (1999). *Vascular plants of Ukraine: A nomenclatural checklist.* (S.L. Mosyakin, Ed.). M. G. Kholodny Institute of Botany, National Academy of Sciences of Ukraine, Kiev.

Parnikoza, I.Y. and Inozemtseva, D.M. (2009). Monitoring of the single natural population of *Cyclamen kuznetzovii* Kotov et Czernowa. *In:* Materials of the IV International Conference of Young Scientists *'Biology: From a molecule to biosphere'* (In Ukrainian). Nove slovo, Kharkiv, pp. 248–249.

Peregrym, M., Vasyliuk, O., Bronskov, O., Bronskova, O., Shyriaieva, D., Spinova, Y., … Oskyrko, O. (2017). *50 rare plants of Donetsk region* (In Ukrainian). LAT & K, Kyiv.

Peregrym, M., Vasyliuk, O., Shyriaieva, D. and Kolomytsev, G. (2014). *50 rare plants of Luhansk region* (In Ukrainian). Veselka, Kyiv.

Shelyag-Sosonko, Y.R. (Ed.) (1996). *The Red Data Book of Ukraine. Plant Kingdom* (In Ukrainian). Ukrains'ka entsyklopedia, Kyiv.

Shevshenko, O.S., Inozemtseva, D.M., Mironova, L.P. and Parnikoza, I.Y. (2008). On the details of *Cyclamen kuznetzovii* Kotov et Czernova miirmecochory in natural and artificial habitats. *In:* Materials of the VIII Ukrainian Scientific Conference of Students and Young Researchers (In Ukrainian). *'Biological studies of young scientists in Ukraine'*, Kyiv, pp. 59–60.

Sytnik, K.M. (Ed.) (1980). *The Red Data Book of the Ukrainian SSR* (In Ukrainian). Naukova Dumka, Kyiv.

Takhtajan, A.L. (Ed.) (1975). *The Red Data Book. The wild species of the flora of the USSR which are in need of protection* (In Russian). Nauka, Leningrad.

Vasyliuk, O., Kuzemko, A., Spriahailo, O., Spriahailo, O., Chorna, H., Shevchyk, V. and Shyriaieva, D. (2018). *50 rare plants of Cherkasy region* (In Ukrainian). Cherkasy.

Vasylyuk, O.V. and Ilnytska, V.I. (Eds) (2010). *Who were you when you were young? Memoirs of participants in the movement of Student Nature Protection Brigades of Ukraine* (In Ukrainian). National Ecological Center of Ukraine, Kyiv.

Zalesskyi, K. (1915). Falz-Fein's preserve steppe in Askania-Nova (in Russian). *Bulletin of the Kharkov Society of Naturalists*, 5, 17–32.

Chapter 9
Awareness of Biological Diversity and Endangered Plant Species and Ecological Education

Abdel Rahman Al-Tawaha, Mariyana Lyubenova, Kamelia Yotovska, Ali M. Qaisi, Ibrahim Al Rawashdeh, Josef Křeček, and Muhamad Aasim

1 Introduction

The word "Endemic" comes from the Greek word *endemos*. Endemism is the presence of a species in a particular region, which can extend from a few square metres to a continent. However, usually, in practice, the term refers to a native species of limited distribution in a particular geographical area or habitat (Kaya and Aksakal 2005).

Worldwide, biodiversity is declining and current trends could have adverse consequences for humanity. Loss of biodiversity threatens the functioning of ecosystems and, hence, the provision of goods and services to people. Biodiversity and the ecological services it helps provide are being depleted progressively every day by human actions and decisions. To mitigate the impacts of such global environmental problems, global networks of protected areas of conservation importance are being built and international agreements, strategies, policy and normative documents for

A. R. Al-Tawaha (✉) · I. Al Rawashdeh
Department of Biological Sciences, Al Hussein Bin Talal University, Ma'an, Jordan
e-mail: abdel-al-tawaha@ahu.edu.jo

M. Lyubenova · K. Yotovska
Faculty of Biology, St. Kliment Ohridski University of Sofia, Sofia, Bulgaria

A. M. Qaisi
Department of Pharmaceutical Sciences, School of Pharmacy, University of Jordan, Amman, Jordan
e-mail: amqaisi@ju.edu.jo

J. Křeček
Department of Hydrology, Czech Technical University in Prague, Prague, Czech Republic
e-mail: josef.krecek@fsv.cvut.cz

M. Aasim
Department of Biotechnology, Faculty of Science, Necmettin Erbakan University, Konya, Turkey

E. Pénzesné Kónya et al. (eds.), *Environmental Sustainability Education for a Changing World*, https://doi.org/10.1007/978-3-030-66384-1_9

nature protection have been prepared. All of these highlight the importance of environmental education.

Biodiversity is the sum total of all of the plants, animals, fungi, and microorganisms along with the communities and ecosystems of which they are a part (Rozensweig 1995; Gaston and Spicer 2004). The Convention on Biological Diversity (CBD) defines biodiversity as the variability among living organisms from all sources including, inter alia, terrestrial, marine and other aquatic ecosystems and the ecological complexes of which they are part; this includes diversity within species, between species and of ecosystems (CBD 1992). Globally, plant biodiversity is threatened by habitat destruction, over-exploitation, agricultural practices and climatic changes. It is also affected by evolutionary processes and socio-cultural changes, including the loss of related local knowledge (Guarion et al. 1995; Hunter and Gibbs 2007; Hawkes et al. 2000). Species losses, and other declines in biodiversity, outcome from four major causes, namely: (i) direct exploitation; (ii) habitat loss and degradation; (iii) introduced species; and (iv) extinction cascades (Gaston and Spicer 2004). Water deficit is the main natural hazard facing Jordan in addition to potential for periodic earthquakes. The main issues related to the environment in this country comprise the limited natural freshwater resources, deforestation, overgrazing soil erosion, desertification and pollution (IUCN-ROWA 2014).

The Intergovernmental Platform for Biodiversity and Ecosystem Services (IPBES) was established in 2012 to provide a mechanism recognised by both the scientific and policy communities to synthesise, review, assess, critically evaluate and deliver relevant knowledge generated worldwide. Current membership includes 126 governments, along with academia, scientific organisations, non-governmental organisations and indigenous communities. The platform operates under the auspices of UNEP, FAO, UNDP, and UNESCO, administered by UNEP. IPBES Deliverable 2(c) is the Global Assessment on Biodiversity and Ecosystem Services, which addresses the status and trends with regard to biodiversity and ecosystem services and their impact on human well-being, as well as the effectiveness of responses, including the strategic plan and its (Aichi) biodiversity targets.

The purpose of this chapter is to present a brief overview of environmental education about biological diversity and awareness of endangered species and their protection in Bulgaria, Jordan, Turkey and Pakistan (Lyubenova et al. 2017).

2 Awareness of Biological Diversity and Endangered Species in Jordan

The flora of Jordan is rich, diverse. Jordan's 2622 species of vascular plants represent 1% of the world flora. However, Jordan's flora shows an alarming trend of degradation and loss due to land conversion and habitat destruction, over exploitation, overgrazing and other anthropogenic and environmental threats. In particular, these problems affect around 485 species belonging to 330 genera and 99 families that are medicinal, aromatics, herbs and spices (Oran 1994; Al-Esawi 1998). These plants have been used, locally, in folk medicines for human and animal care. They generate income for resource-poor people, especially women, who are involved in most of the collection, processing and marketing of medicinal plants.

Jordan embraces four bio-geographic regions, namely: the Mediterranean, Irano-Turanian, Saharo-Arabian and Sudanian (Al-Esawi 1998). The four regions include 13 vegetation types, which provide the natural habitats for over 4000 species of fauna and flora from the terrestrial, marine and freshwater. Human activities are causing major impacts and so the monitoring of particularly endangered plants is a fundamental requirement for guiding the management and conservation of species and habitat. The number of endangered plant species has increased due to changes in land use, farming, clearing, road construction, buildings, mining, and rural harvesting of edible wild plants such as *Gundelia tournefortii L.* and medicinal plant such as *Artemisia herba alba*, *Teucrium polium* L., *Paronychia argentea*, *Matricaria aurea* (loef). Schultz. Bp. and *Ziziphora tenuior*. Pastoral communities are well aware about the use of plants for treating the ailments of animals, such as: *Teucrium polium*, *Peganum harmala*, *Hypericum* spp., *Retama raetam* and *Varthemiai phionoides* (IUCN-ROWA 2014). However, *Gundelia tournefortii* L. is considered as a cash crop that most pastoralists collect and thus, it has quickly become endangered. The loss of habitat is the prevalent driver of extinction, and it is important to understand current risk (Dirzo and Raven 2003). In order for medicinal wild plants to survive in their natural ecosystems, suitable conservation practices must be adopted and awareness created among the people who inhabit these lands (Roa et al. 2014).

The IUCN species conservation programme and commission have been assessing the conservation status of species, subspecies, varieties, and selected subpopulations on a global scale for the past 50 years in order to highlight taxa threatened with extinction and to promote their conservation at all levels. The Jordan's Royal Botanic Garden (RBG) has led the assessment of Jordanian plants including more than 1000 plant species. The Red List of the Jordanian Flora aims to support the creation of conservation strategies that will both help protect Jordan's plant diversity and serve as a reference for future studies and monitoring programmes.

Increased public awareness is critical for biodiversity conservation and sustainable land use. Public awareness about the endangered plants of Jordan was initiated through the media, farm visit, interviews, schools and courses on biodiversity included in biology curricula at Jordan universities. Mass media, workshops and school activities were used to introduce the importance of dryland agro biodiversity to a large public audience (Abbass 2001). At the level of schools, painting competitions were launched to attract children who had an interest in conserving the environment and biodiversity. A collaboration with a rural theater team also helped expand awareness of biodiversity and the conservation of endangered plants.

For the future: regional as well international cooperation programmes for conservation and protection of the endangered and endemic plants are needed for the exchange of germplasm between countries to help minimize the loss of targets species and also restore them when conditions are suitable, especially in countries exposed to wars. Endangered plants can be conserved at a gene bank for long time (−70 °C) through collection of their seeds or cuttings. Governments can also help by promoting and establishing nature reserves and then monitoring them to keep the species from outside influences. Biosafety guidelines and biodiversity action plans are essential within countries' legislation. The conservation and sustainable use of the important medicinal plants in Jordan's nature reserves requires strict regulations for both the grazing and the collection of plants. The National Biodiversity Strategy and Action Plan (NBSAP) proposed a series of priority objectives and actions that are presented according to five themes involving most sectors of society. These include: (i) protection of biological resources; includes endangered species; national red list; and protected areas, and (ii) building a biodiversity-oriented society, which includes economic valuation of biodiversity; legislation, institutional structures, public awareness and participation.

3 Awareness of Biological Diversity and Endangered Species in Bulgaria

Bulgaria has a well-developed network of protected areas that cover nearly 5% of the country's territory. From Bulgaria's Protected Areas Act, one management objective for national and natural parks, reserves, maintained reserves and protected areas is carrying out educational activities in the field of ecology and environmental protection (Protected Areas Act 2007).

Of course, education is an important element of nature protection and of global importance. Commitments on this issue have been included since the earliest international conventions, such as the: Convention on the Protection of the World Cultural and Natural Heritage (1972), Convention on the Conservation of Wild European Flora and Fauna and Natural Habitats (1979) and Convention on Biological Diversity (1993).

The necessity for environmental education has been recognized in Bulgaria since the adoption of the Ordinance on the Conservation of Nature (1936) with the active cooperation of the Union for the Protection of the Homeland (1929) and the nature conservation societies and organisations included therein. The normative legal documents including Nature Protection Act (1967), Protected Areas Act (1998), continue this tradition and educational activities remain an integral part of the management objectives of protected areas in Bulgaria.

Education for Sustainable Development is a next stage in the development of eco-education, where environmental issues become equally important with socio-economic development issues. Formally, it has emerged since the United Nations Conference on Earth Issues (1992). Its subsequent implementation is set in the: Agenda 21 for Action on Sustainable Development of the Planet (1992), Johannesburg Plan for the Implementation of Agenda 21 (2002) and Global Plan of Action for Education for Sustainable Development (2014).

Protected areas play an important role as natural learning laboratories in achieving the goals of sustainable development education. They foster appropriate education through formal education, extracurricular activities and non-formal education that promotes knowledge in the field of ecology and environmental protection. This learning approach, which applies active and experiential learning and uses nature as an educational tool, is thought effective.

Non-formal education became part of the international discourse on education policy in the late 1960s and early 1970s. It can be seen to be related to the concepts of recurrent and lifelong learning. Non-formal education is any organised educational activity outside the established formal system, whether operating separately or as an important feature of some broader activity, that serves identifiable learning clienteles and learning objectives (Coombs and Ahmed 1974).

Bulgarian biota ranks among the most species-rich in Europe, especially relative to the country's size. More than 3100 *algae* species from 580 genera, 15 classes and seven divisions have been discovered so far. The *bryoflora* contains 754 species, of which 34% are of conservation importance (Critically Endangered, CR – 18; Endangered, EN – 25; Vulnerable, VU – 124 and Near Threatened, NT – 20). *Pteridophytes and seed plants* (incl. neophytes) are 4100 species of which 270 are Balkan, and 174 are Bulgarian, endemics (Golemansky 2015). As Erika p-Konya advises list 801 Red-list species, of which: Extinct, EX – 1; Regionally Extinct, RE – 12; CR – 208, EN – 297, VU – 204 and NT – 79. Overall, this constitutes 19.5% of the national flora. So far, 340 neophytes have been registered in Bulgaria, 16 in the period 2009–2013. More than 770 Bulgarian species (around 21% of the flora) are medicinal plants. The number of plants used frequently in traditional and official medicine is 220 species.

Animals recorded in Bulgaria amount to 30,359 species from 28 types and 75 classes but this is only 50% of the estimated total animal diversity. The list includes 27,000 invertebrates (about 85% are insects). There are 75 vertebrate species, the best studied group, which include 97 species of mammals, 409 birds, 37 reptiles, 19 amphibians, and 219 Black Sea and fresh water fish. Bulgaria is home to 33 of Europe's 35 species of bat. The large mammal fauna, with 23 species, is also among the richest in Europe. It includes the wolf (*Canis lupus lupus*), brown bear (*Ursus arctos*), badger (*Meles meles*), two species of marten (*Martes martes* and *M. foina*), three species of polecat (*Mustela evarsmanni, M. purtorius* and *Vormela peregusna peregusna*), three species of deer (*Dama dama, Cervus elaphus* and *Capreolus capreolus*), and three marine mammals (*Dolphinus delphis, Phocaena phocaena relicta* and *Tursiops truncates ponticus*). A fourth marine mammal, the Black Seamonk seal, a local form of the Mediterranean monk seal (*Monachus monachus*), is presumed extinct. The Bulgarian herpeto-fauna ranks third in Europe in species richness after Spain and Greece. There are 1200 endemic animal species (790 Bulgarian and 410 Balkan endemics). More than 780 cave species occur in Bulgaria, of which 110 are typical troglobionts.

For the conservation of this biodiversity, the '*Red Data Book of Republic of Bulgaria*' has been published in three volumes: for plants and fungi, for animals, and for natural habitats. Organisms are divided into several categories depending on their status: disappeared, threatened, vulnerable, etc. The book presents the need for regeneration and restoration of the Bulgarian flora and fauna. As elsewhere, point of this initiative is to increase public awareness of the state of flora and fauna. For example, the latest edition of the *Red Data Book of Republic of Bulgaria's Volume. 2: Animals*, lists 442 species: EX (Extinct) – 30, CR (Critical) – 87, EN (Endangered) – 107, VU (Vulnerable) – 137, NT (Not threatened) – 14, Least Concern (LC) – 42 and Data (DD) – 25. Fungi and fungus-like species are much less studied but, so far, over 4900 species have been recorded from a predicted total of 20,670. In the last 5 years, five volumes of '*Fungi in Bulgaria*' have been published with original data for more than 470 species. The Red List of fungi in Bulgaria contains 215 species of ascomycetes and basidiomycetes, which are: CR – 37, EN – 105, VU – 40, NT – 14 and DD – 19 (Peev 2015; Lyubenova et al. 2017).

Specific work related to biodiversity conservation is guided by the National Strategy for Biodiversity Conservation (NSBC 1998). Bulgarian legislation for biological diversity includes the following Acts: the Environment Protection Act (2002), the Biological Diversity Act (2002), the Medicinal Plants Act (2000), the Forestry Act (2011), the Fisheries and Aquaculture Act (2005), the Hunting and Game Protection Act (2000), the Genetically Modified Organisms Act (2005) and so on. The Bulgarian ecological legislation is fully harmonised with European legislation. According to the Bulgarian Biological Diversity Act, in Bulgaria, there are legally protected: 1057 endangered species: 595 species of vascular plants and 443 species of animals.

The Medicinal Plants Act regulates activities concerning the preservation and use of medicinal plants, the liabilities of physical persons and legal entities and governmental and local authorities for conservation of biological diversity and resources of medicinal plants. The general purpose of this act is to provide conditions for the long-term sustainable use of medicinal plants across the whole territory. About 61 medical species are protected by the Biological Diversity Act, which introduces a system of protected areas with six categories under International Union for Conservation of Nature (IUCN) and defines the interactions between the institutions responsible for their management.

At the end of 2016, Bulgaria had 1012 legally designated protected areas in six categories (National Parks – 3, Natural Parks – 11, Strong Reserves – 55, Managed Reserves – 35, Protected Places – 564 and Landmarks – 344); that is 5.2% of country's territory. The law has also imposed management plans. The Natura 2000 Network in Bulgaria covers 4,105,320 ha or 34.4% of the territory. This network includes 119 Special Protection Areas (22.7% of the territory) and 233 Special Areas of Conservation (30% of the whole territory). By the share of the national land area given to National Natura 2000 Network, Bulgaria ranks third in the European Union after Slovenia (35.52%) and Croatia (34.80%). The National Ecological Network comprises sites of the European Natura 2000 sites (Special Areas of Conservation under the Habitats Directive and Special Protection Areas under the Birds Directive) and protected areas which may be included in or may be outside the Natura 2000 area (Lyubenova et al. 2017).

Bulgarian environmental education is controlled by the standards set for the secondary school curriculum, which are seen as a mechanism for management and regulation of the general education system. They propose a normative curricula for each respective class. This curriculum decomposes the goals (the expected results), in terms of a "core of learning content – knowledge (concepts, laws, theories), skills, relationships" (Tsanova 2007).

For example, Chap. 3 (paragraph 22) of the Pre-school and School Education Act regulates standards for civic, health, environmental and intercultural education (Pre-school and School Education Act 2016). An important step towards regulating ecological education in Bulgaria is Ordinance No. 13 on Civil, Health, Ecological and Intercultural Education. In this normative document, environmental education is defined as: "aimed at the formation of ecological culture, ecological awareness and ecological behaviour in their mutual relation with a view to knowledge of environmental laws, protection, improvement, management and reasonable use of natural resources as well as protection of the natural environment and ecological equilibrium". Ordinance No. 13 regulates the framework requirements for the results of education in the field of environmental education for pre-school and school education – by degrees and stages. Expected results are in areas of competence: Water, Soil, Air, Energy and Climate, Biodiversity, Society and the Environment (Ordinance No. 13 2016). In Ordinance No. 5 of 30 November 2015 on general education, paragraph 3 (3) reads: "Supporting sustainable development principles, including ecological, economic and social dimensions and aiming at outlining a long-term vision for society, is through training on different subjects and aims to build a personal responsibility for sustainable future." (Ordinance No. 5 on general education 2015; Ordinance No. 5 on Pre-school Education 2016). The Comparative Review of the Standards for Different Degrees and Stages regarding the opportunities for environmental education of students represents this as a set of knowledge, skills and attitudes to be develped within the general education of students (Table 9.1).

The conclusions drawn from this analysis of the curriculum to the end of the first high school in the context of environmental culture, environmental awareness and environmental behaviour are:

- The centre is knowledge of environmental law, protection, improvement, management and reasonable use of natural resources;
- The central concern is the preservation of the natural environment and the ecological balance;
- Science knowledge is used to formulate a problem (including environmental), acquire new knowledge, explain natural processes and phenomena, make substantiated conclusions on issues related to natural sciences;
- The features of natural sciences as part of human knowledge include nature study as a major feature;
- Natural sciences and technologies form in a certain way our material, intellectual and cultural environment;

Table 9.1 Analysis of the integration place for the endangered plant species knowledge in Bulgaria of the students by degrees and stages

Level of education	Knowledge integration possibilities for endangered plant species	Class
Basic level of education, initial stage	**Environment** Areas of competence: Natural diversity The man and the healthy Lifestyle	3–4 class
	The man and the society Area of competence: The man and his Environment	3–4 class
Basic level, lower secondary level	**Biology and health education** Area of competence: Organism-environment	7 class
Secondary education, first high school stage	**Biology and health education** Areas of competence: Multicellular organism Biosphere Observations, experiments, research	10 class
Secondary education, second high school stage	**Biology and health education**	12 class

- The scientific explanations, the identification of scientific problems, the scientific explanation of natural processes and phenomena, the use of scientific data and evidence can also be applied in an ecological context;
- The centre is to discuss issues related to natural sciences and sharing ideas; and
- The appearance is individual, public and global: natural resources, environment, risk.

Unfortunately, expected results about endangered species are lost in this educational process. For example, there are only two themes (one in the 7th and one in the 10 class) of current content where endangered plant species are mentioned. In the 7th class, knowledge of the rules of conduct in protected areas are presented, which does address endangered species.

However, in addition to the standard class-lesson forms of organising the learning process, extracurricular activities are also used successfully. During excursions, students experience natural objects and endangered plant species, perceiving them in their entirety, acquiring broad experiences related to nature, collecting rich visual material, which they can then use in the classroom (Asenova and Yotovska 2014). Observing different ecosystems and their great diversity of species leads to a better understanding. Students recognize the need to preserve the environment and to preserve its integrity. Outdoor work (including eco-paths) or other projects organised by various nature conservation organisations also contribute to this learning. Many learning tools have been created that can be applied successfully in formal education (Black Sea box, Danube package, etc.; Asenova et al. 2012; Miteva et al. 2009, 2012; Yotovska et al. 2016). Some include knowledge about protected plants and other endemic species. By discussing life problems and their involvement in different projects, a positive and proactive attitude towards the environment may be formed. As for non-formal education in sustainable development, some serious work with an ecological focus being done on the development of: training programmes of different duration in protected areas. These are being prepared in a logical relationship with each other and are tailored to the participants' age so providing tools for non-formal education for sustainable development; and mechanisms for evaluating the quality of educational activities (Yotovska et al. 2015; Bancheva et al. 2016).

Today, according to state requirements, the curricula of almost all bachelor's degree programmes in Bulgarian universities include "Ecology and Environmental Protection" as a general-purpose course. This also provides information on the ecosystem role of biodiversity, its status and conservation, and the types of threats to species (Table 9.2).

Student enrolment is higher for universities offering the specialty 'Ecology and Environmental Protection'. A Faculty of Ecology and Environmental Protection has been established at the Forestry University, while in other universities, such as the Sofia University, there are separate departments. For example: Department of Landscape Ecology and Nature Conservation at the Geography and Geology Faculty; and Ecology and Environmental Protection in the Faculty of Biology. There are also Master's courses in the faculty of Biology such as "Biological Resources and their Protection", "Biodiversity and Conservation", "Environmental Legislation", "Environmental Impact Assessment", etc.

A Master's programme in Ecological Education has been active in the Faculty of Biology, Sofia University "Kl. Ohridski" for 15 years. This "Pedagogue-Master of Environmental Education" is interdisciplinary and provides specialised training in both ecology and education. This Master's degree contributes to the general professional development of teachers and priority attention is paid to the problems of ecology, ecological education and upbringing. The qualification of graduates of this Masters programme allows them to carry out specialised activities in all types of secondary schools (compulsory and extra-curricular), as well as various nature conservation organisations, reserves, visitor centres, and so on.

Table 9.2 Analysis of the bachelor's degree programmes in Bulgaria's 51 accredited universities

Indicator	Bulgarian universities
Bachelor's degree programmes "ecology"	21
Bachelor's degree programmes in Bulgarian universities include "ecology and environmental protection"	43

4 Awareness of Biological Diversity and Endangered Species in Pakistan

Pakistan is a diverse geographical region ranging from Arabian sea in the south to its northern mountains, which have peaks above 8000 m from sea level. Due to its diverse biomes, Pakistan has rich biodiversity which includes 180 mammals, 700 birds and 540 fish species (Anonymous 2019b). Unfortunately, massive destruction of habitat due to the cutting of forest for energy consumption, furniture making and urbanisation has adversely affected Pakistan's endemic species.

Pakistan is a land of over 5700 diverse plant species. Out of these, 405 plants are endemic plant species from 43 families (Anonymous 2011). According to a 2011 report published in Dawn News (a well-known newspaper of Pakistan), 703 species are endangered (Anonymous 2011). This report also highlighted the plant species that no longer exist in Pakistan. These plants include: *Allium gilgiticum* (Gilgit), *Arabidopsis brevicaulis* (Hunza valley), *Asparagus gharoensis* (Sindh), *Bruguiera gymnorrhiza* (Indus delta), *Cousinia matifeldei* (Chitral), *Nepeta schinidii* (Chitral), *Pedicularis caeruleoalbescens* (Chitral), *Saxifraga duthei* (Baltistan), *Scaveola plumererii* (Sindh coast), *Scaveola taccada* (Sindh coast), *Sonneratia caseolaris* (Indus delta) and *Taraxacum chitralicum* (Chitral).

According to the IUCN's *Red List*, 18 plant species are globally endangered (Anonymous 2019c), while 12 higher plant species were reported threatened by the World Bank's 2016 collection of development indicators (Anonymous 2019d). Pakistan is one of those countries where people know very little about endemic and endangered species or their ecosystem role. Unfortunately, research on endangered plants by Pakistan's scientific community is also very limited and, as the IUCN notes, little information is available about the endangered plant species of Pakistan. However, field studies of *Astragalus gilgitensis* (Alam and Ali, 2009) and *Cadaba heterotricha* Stocks (Abbas et al. 2010) place them in the Critically Endangered category and Endangered categories respectively.

Alam and Ali (2010) also indicate the number of endangered plant species in Pakistan but only for 8 of 19 taxa. 19 flowering plants were awarded IUCN Red List status at national (Anonymous 2001) and regional level (Anonymous 2003) using the IUCN Red List's categories and criteria (Anonymous 2008). Of this 19, 16 were from Gilgit and Baltistan (Alam 2009), one was endemic to Sind Province (Ali and Khan 2009), and two were restricted to the coastal area of Karachi (Ali 1972). Alam and Ali's (2010) work categorizes the data as follows:

(I) *Collected Threatened Taxa*

  (a) Critically Endangered (CR) Taxa

    (i) *Tanacetum baltistanicum*
    (ii) *(ii) Haplophyllum gilesii*
    (iii) *Berberis pseudumbellata* subsp. *gilgitica*
    (iv) *Astragalus clarkeanus*

(v) *Asperula oppositifolia* subsp. *baltistanica*
(vi) *Androsace russellii*

(b) Vulnerable (VU) Taxa

(i) *Rhodiola saxifragoides*
(ii) *Aconitum violaceum var. weileri,*

(II) *Uncollected Taxa*

(a) Extinct (EX) taxon

(i) *Asparagus gharoensis* Blatter:

(b) Regionally extinct (RE) taxa

(i) *Scaevola taccada* (Gaertn.) Roxb
(ii) *Scaevola plumierii* (L.) Vahl

(c) Possibly extinct

(i) *Saxifraga duthiei* Gandogar
(ii) *Plantago baltistanica* Hartmann
(iii) *Mattiastrum karakoricum* Podlech and Sadat,
(iv) *Elymus russellii* (Meld.) T. A. Cope,
(v) *Consolida schlagintweitii* (Huth) Munz,
(vi) *Christolea mirabilis* (Pamp.) Jafri,
(vii) *Arabidopsis brevicaulis* (Jafri) Jafri,
(viii) *Allium gilgiticum* Wang and Tang

Hussain et al. (2010) reported on the endangered plants of Karachi, the largest city of Pakistan at the coast of Arabian Sea. Their report recognises: 18 threatened tree species, 11 threatened shrub species, 29 threatened under shrub species, 11 threatened woody climbers, seven threatened herbaceous climber species and 59 threatened herbaceous plants, a total of 135 threatened plant species. These plants, classified as vulnerable, rare, extinct, endangered, invasive and intermediate, are listed in Table 9.3.

Table 9.3 Distribution of different endangered plants of Karachi, Pakistan

	Total	Vulnerable	Rare	Extinct	Endangered	Invasive	Inter-mediate
Tree	18	12	2	1	1	2	0
Shrubs	11	6	4	0	1	0	0
Under shrub (threatened)	29	19	7	3	0	0	0
Woody climbers	11	3	6	2	0	0	0
Hebaceous climber	7	4	2	1	0	0	0
Herbaceous plants	59	22	25	6	3	0	2
Total	135	66	46	13	5	2	2

Haq (2011) has reported on the critically endangered flora of District Battagram, Pakistan, finding 12 critically endangered plant species. Species suffering population decrease (%) were: (1) *Acer caesium* Wall (86%), (2) *Betula utilis* D. Don (98%), (3) *Cedrus deodara* Roxb. ex Lamb (98%), (4) *Opuntia dilleni* Haw (85%), (5) *Paeonia emodi* Wall. ex Hook. f (81%), (6) *Pistacea integerrima* (J. L. Stewart) Rech.f (93%), (7) *Populus alba* Linn, (8) *Quercus glauca* Thunb, (9) *Skimmia laureola* D.C (81%), (10) *Taxus baccata* L. Subsp. *Wallichiana* (Zucc.) Pilger (87%), (11) *Ulmus wallichiana* Planch and (12) *Viscum album* Linn. These, often significant, decrease in plant population was caused by habitat loss including area of occupancy, human exploitation of resources, the introduction of exotic taxa and pathogens (Haq 2011).

Pakistan has diverse topography and climatic conditions which create diverse ecosystems and habitats. Four of these are recognised as part of the Global 2000 list of most significant ecosystems. Unfortunately, Pakistan has no agency for ecological conservation and no action plan for landscape and land use; the consequence will likely be further ecological degradation and loss of biodiversity. Pakistan ranks among the top 10 countries that face severe biodiversity damage due to climate change and some animals and plant species are at risk of extinction. Meanwhile, the increasing cultivation of marginal lands for crops or animal production is also causing a threat to the wild flora. This is fuelled by factors that include rapid population growth, deforestation for fuel, wood, forage, illegal trade, over grazing, erosion, water logging, etc., which are causing major losses to specific habitats and, in turn, increased danger of habitat fragmentation and eventual extinction of species. Erosion of the genetic diversity of major crops is also very high in Pakistan.

The Climate Change Vulnerability Index 1 ranks Pakistan 16th in 2014 and thus one of the countries most affected by climatic changes, which result in 14 billion $ loss per annum. Pakistan is always looking for sustainable development programmes and, in 1992, a National Conservation Strategy was developed at the request of the United Nation's Convention on Biological Diversity (UN-CBD). Such policies were also developed at Provincial and District levels. The threat of extinction to major parts of its flora and fauna forced Pakistan to sign and cooperate with the global community's environment related treaties (Anonymous 2015). These include: (i) Convention on Biological Diversity (CBD), (ii) Convention on the Conservation of Migratory Species (CMS), (iii) Convention on International Trade in Endangered Species of Wild Flora and Fauna (CITES), (iv) Cartagena Protocol on Biosafety, (v) Nagoya Protocol on Access and Benefit Sharing, (vi) The United Nations Framework Convention on Climate Change (UNFCCC), and (vii) the Sustainable Development Goals.

The basis for Pakistan's national action plans for the conservation of biodiversity is Article 9 of the Constitution of Pakistan, which highlights the "right to life" for biodiversity. The Article's 18th amendment devolves biodiversity conservation to Provincial Governments but the Federal Government remains empowered to make International treaties. Other acts like Pakistan Environmental Protection Act (1997), Pakistan Trade Control of Wild Fauna and Flora Act, 2012 (CITES Act) and Access Benefit Sharing Bill (2012) have also been constructed. Similarly, a number of

"Biodiversity Policy Frameworks" have been launched and included in the National Conservation Strategy (NCS – 1992), Biodiversity Action Plan (BAP – 2000), National Environment Policy (2005), National Sustainable Development Strategy, 2012 (NSDS) and National Climate Change Policy (2012). All these programmes aim to promote sustainability and create awareness among people. However, political and economic instability has resulted in unsatisfactory progress on the Biodiversity Action Plan (BAP), 2000.

The BAP 2000 programme was formulated by a Biodiversity Working Group (BWG) which had representatives from government, civil society and experts of biodiversity and was approved by Pakistan's Environmental Protection Council in 2000. Thereafter, a roadmap for BAP was overseen by BWG and a programme was launched with the help of provincial governments. This programme addresses: biodiversity awareness, gender, poverty, the biodiversity nexus, and also mainstream biodiversity concerns into national planning and policy processes for terrestrial ecosystems, individual habitats, species, forest ecosystems, inland wetland ecosystems, coastal and marine ecosystems, sustainable agriculture and agro-biodiversity, sustainable production and consumption and biosafety issues and challenges.

During implementation, different stakeholders found it difficult to understand the concepts of biodiversity and its values. Therefore, a Biodiversity Directorate was established which suggested and launched thematic workshops to develop knowledge about biodiversity, increase understanding of the national BAP and develop training for using relevant tools and mechanisms. The main features and targets for 2020 of this programme are:

1. Creating awareness among people about biodiversity, conservation and sustainability
2. Integrating biodiversity values with poverty reduction and planning for national and local development
3. Integrating different stakeholders for sustainabile resources management
4. Reducing the rate of biodiversity loss to near zero for all natural habitats
5. Promoting sustainable land use for areas under agriculture, aquaculture and forestry
6. Reducing pollution
7. Identifying invasive alien species and their pathways to prevent their introduction and establishment
8. Preventing the extinction of known threatened species
9. Maintaining the genetic diversity of cultivated plants and domesticated animals and their wild relatives to minimise genetic erosion
10. Enhancing and implementing relevant traditional knowledge, innovations and the practices of indigenous and local communities relevant to conservation and sustainable biodiversity
11. Applying scientific research and innovative technologies to biodiversity protection.

After implementing the national BAP programme, the following plant species have been targeted for conservation (Anonymous 2015).

1. Aconite, Atees (*Aconitum heterophyllum*) AJK, KP, GB
2. Alder (*Sharol Alnusnitida*) AJK, KP
3. Bakau (*Rhizophora mucronate*) Sindh
4. Chilghoza (*Pinus gerardiana*) Balochistan, GB
5. Costus, Kuth (*Saussurea costus*) AJK, KP, GB
6. Gugul (*Commiphora wightii*) Sindh
7. Himalayan May apple (*Podophyllum hexandrum*) AJK, KP, GB
8. Himalayan Yew (*Taxus wallichiana*) AJK, KP, GB
9. Indian Maple (*Acer caesium*) AJK, KP
10. Indian snakeroot (*Rauvolfia serpentine*) AJK, KP
11. Juniper (*Juniperus excelsia*) Balochistan
12. Kutki (*Picrorhiza kurroa*) AJK, KP, GB
13. Marsh orchid or Spotted Orchid (*Dactylorhiza graggeriana*) AJK, GB
14. Marsh orchid or Spotted Orchid (*Dactylorhiza hatagirea*) AJK, GB
15. Mazri (*Nannorrhops ritchiana*) Balochistan, KP, Punjab
16. Nag Chhatri (*Trillium govanianum*) AJK, Gb
17. Nepalese Alder (*Alnus nepalensis*) AJK, KP
18. *Primrose Primula* sp. AJK, KP, GB

Field results clearly reveal that plant species in Pakistan are under threat of extinction due to several socio-economic and environmental factors. The number of endangered species in Pakistan differs in data taken from various sources. Just 18 plants are listed by the IUCN; this clearly highlights the problem as regional studies reveal much higher number of endangered species in Pakistan. There is a dire need to control the endemic and endangered plant species of all regions followed by enlisting the plants to IUCN data base for better conservation. Recently, Pakistan launched its long term "*National Biodiversity Strategy and Action Plan for achieving Aichi Biodiversity Targets and Sustainable Development Goals (2017-2030)*" (Anonymous 2019e). This headlines the need for environmental education to raise public awareness of the biodiversity issues and a process of education has now begun for those directed to implement the plan.

5 Conclusion

This chapter dealt with the status of biological diversity in Jordan, Bulgaria, and Pakistan, its protection and also public education about their endangered species. The flora of Jordan, Bulgaria, and Pakistan is rich, diverse and includes many medicinal and aromatic plants as well as herbs and spices. Information about the biodiversity of Jordan, Bulgaria, and Pakistan as well as about its protection is (or should be) included at all levels of formal education – basic, secondary, higher, informal and non-formal, although the process is in its very early stages in some of these countries. In higher education, the focus of attention is the ecosystem role of biodiversity and its importance for the stability of the biosphere and effectiveness of ecosystem functions, and its role in the provision and regulation of ecosystem service. The main goals of environmental education are consciousness raising and the development of ecological thinking.

Significant biological diversity at all levels – genetic, species and ecosystem, is considered valuable asset of Jordan, Bulgaria, and Pakistan. Biological diversity protection in these countries helps guarantee the preservation of ecosystem integrity and sustainable ecosystem functioning, providing sustainable ecosystem services to society. However, achieving this requires the creativity of environmentally responsible individuals, people who may be made empowered through active environmental education. This extremely important role for education can be realised through the improvement of the formal education courses and increased participation of different forms of non-formal education and voluntary activity that allows learners to gain hands-on experience in solving the real problems and cases.

References

Abbas, D. (2001). Dry land agrobiology: Conservation and sustainable use of dry land agro biodiversity. Unpublished project funded by GEF/UNDP, Syria-Jordan-Lebanon-Palestinian Authority. GEF/UNDP/ICARDA/ACSAD. October-December 2001.

Abbas, H., Qaiser, M. and Alam, J. (2010). Conservation status of Cada heterotricha Stocks (Capparidaceae): An endangered species in Pakistan. *Pakistan Journal of Botany*, 42(1), 35–46.

Alam, J. (2009). *Endemic Flora of Gilgit and Baltistan and Conservation Strategies for Threatened Endemic Taxa.* University of Karachi, Ph. D Thesis, Karachi.

Alam, J. and Ali, S.I. (2009). Conservation status of Astragalus gilgitensis Ali (Fabaceae): A critically endangered species in Gilgit District, Pakistan. *Phyton* (Horn, Austria), 48, 211–223.

Alam, J. and Ali, S.I. (2010). Contribution to the red list of the plants of Pakistan. *Pakistan Journal of Botany*, 42(5), 2967–2971.

Al-Esawi, D.M.H. (1998). *Field Guide to Wild Flower of Jordan and Neighboring Countries.* Al-Rai Publishing, Amman, Jordan.

Ali, S.I. (1972). Goodeniaceae. *In:* Nasir, E., Ali, S.I. (eds), *Flora of Pakistan.* No. 30, Karachi.

Ali, S.I. and Khan S.W. (2009). Asparagaceae. *In:* Ali, S.I., Qaiser, M. (eds), *Flora of Pakistan*, Karachi, Pakistan.

Anonymous (2001). IUCN Red List Categories and Criteria: Version 3.1. IUCN Species Survival Commission. IUCN, Gland, Switzerland.

Anonymous (2003). Guidelines for Application of IUCN Red List Criteria at Regional Level: Version 3.0. IUCN Species Survival Commission. IUCN, Gland, Switzerland.

Anonymous (2008). Standards and Petitions Working Group. Guidelines for the IUCN Red List Categories and Criteria: Version 7.0. Standards and Petitions Working Group, IUCN SSC Biodiversity Assessments Sub-Committee, Gland, Switzerland.

Anonymous (2011). Many local plant species becoming extinct: Experts. *Dawn*, 17, 1.2011, https://www.dawn.com/news/599390.

Anonymous (2015). Pakistan National Biodiversity Strategy and Action Plan (NBSAP), https://www.iucn.org/sites/dev/files/import/downloads/nbsap_1st_draft_23_3_15.pdf.

Anonymous (2019b). *Endangered Species in Pakistan.* Pakistan Environmental Journalists Forum. http://envjournalists.com/endangered-species-in-pakistan/.

Anonymous (2019c). *Endangered Plants in Pakistan.* Pakistan Environmental Journalists Forum. http://envjournalists.com/endangered-species-in-pakistan/endangered-plants-in-pakistan/.

Anonymous (2019d). *Pakistan: Plant species (higher), threatened*. Trading Economics, https://tradingeconomics.com/pakistan/plant-species-higher-threatened-wb-data.html.

Anonymous (2019e). *Pakistan National Biodiversity Strategy and Action Plan for achieving Aichi Biodiversity Targets and Sustainable Development Goals*. CBD International with Government of Pakistan, Islamabad, https://www.cbd.int/doc/world/pk/pk-nbsap-v2-en.pdf.

Asenova, A. and Yotovska, K. (2014). Extracurricular activities in secondary school – analysis of the state and prospects. *Scientific Papers of the University of* Rousse, 53(11), 102–106.

Asenova, A., Yotovska, K. and Evtimova, S. (2012). Project based training on the topic: Biodiversity in Bulgarian symbols and traditions. *International Scientific Journal: Science & Technologies* (Stara Zagora), 2(8), 14–18.

Bancheva, H., Yotovska, K. and Asenova, A. (2016). Methodology for assessing and impacting the work of organizations in the field of Education for Sustainable Development. *Management and Sustainable Development, University of* Technology, 4(59), 45–50.

CBD (1992). *Text of the Convention on Biological Diversity*. CBD website: http://www.cbd.int/convention/convention.shtml.

Coombs, P. and Ahmed, M. (1974). *Attacking Rural Poverty*. John Hopkins University Press. Baltimore.

Dirzo, R. and Raven, R.H. (2003). Global state of biodiversity and loss. *Annu. Rev. Environ. Resour.*, 28, 137–167.

Gaston, K.J. and John I. Spicer (2004). *Biodiversity: An introduction* (2e). Blackwell.

Golemansky, V. (Ed.) (2015). *Red Data Book of Republic of Bulgaria, Vol. 2. Animals*. Bulgarian Academy of Sciences, Ministry of Environment and Water, Sofia.

Guarion, L., Rao, V.R. and Reid, R. (Eds) (1995). *Collecting Plant Genetic Diversity: Technical Guidelines*. CAB International, Wallingford, UK.

Haq, F. (2011). The Critically Endangered flora and fauna of District Battagram, Pakistan. *Advances in Life* Sciences, 2, 118–123.

Hawkes, J.G., Maxted, N. and Ford-Lloyd, B.V. (2000). *The Ex Situ Conservation of Plant Genetic Resources*. Kluwer Academic Publishers, Netherlands.

Hunter, M.L. Jr. and Gibbs, J.P. (2007). *Fundamentals of Conservation Biology* (3e). Oxford, Blackwell.

Hussain, S.S., Ahmed, M., Siddiqui, M.F. and Wahab, M. (2010). Threatened and endangered native plants of Karachi. *International Journal of Biology and Biotechnology*, 7(3): 259–266.

International Union for the Conservation of Nature – Regional Office for West Asia (IUCN-ROWA) (2014). *Jordanian Fifth National Report on the Implementation of the Convention on Biological Diversity*. Ministry of Environment, Amman, Jordan, https://www.cbd.int/doc/world/jo/jo-nr-05-en.pdf.

Kaya, Y. and Aksakal, Ö. (2005). Dıstrıbutıon of endemıc plants in the World and Turkey, *Journal of Erzincan Faculty of Education*, 7(1), 85–99.

Lyubenova, M., Stoyanova, K. and Ivanova, C. 2017. Biological Diversity, Ecosystem services and their payment in Bulgaria. *Journal of Balkan Ecology*, 20(3), 229–235.

Miteva, T., Assenova, A. and Drobenov, V. (2009). *Osogovo, Belasitsa and Ograjden – Children's Educational Tool Of 4-8 Grade*. REC, Sofia, Bulgaria [In Bulgarian].

Miteva, T., Asenova, A. and Simov, N. (2012). *Pirin for Children – Educational Tool of 4-8 Grade* [In Bulgarian]. REC, Sofia.

NSBC (1998). *National Strategy for Biological Diversity Conservation*. Ministry of Environment and Water (MEW), Sofia.

Oran, S. (1994). Genetic resources of medicinal plants in Jordan. *In:* A.A. Jaradat (Ed.), *Plant Genetic Resources of Jordan*. Proceedings of a National Seminar: 2-4 August 1994, Amman, Jordan. International Plant Genetic Resources Institute, Regional Office for West Asia and North Africa, Aleppo, Syria, pp. 117–125.

Ordinance No. 13 of 21.09.2016 on *Civil, Health, Ecological and Intercultural Education*, Prom. – SG, No. 80 of 11.10.2016, in force as of 11.10.2016. [In Bulgarian: Наредба №

13 от 21.09.2016 г. за гражданското, здравното, екологичното и интеркултурното образование]. Government of Bulgaria, Sofia.

Ordinance No. 5 of 03.06.2016 on *Pre-school Education*, Prom. – SG, No. 46 of 17.06.2016, in force as of 01.08.2016. [In Bulgarian: НАРЕДБА № 5 от 03.06.2016 г. за предучилищното образование, Обн. – ДВ, бр. 46 от 17.06.2016 г., в сила от 01.08.2016 г.]. Government of Bulgaria, Sofia.

Ordinance No. 5 of 30.11.2015 on *General Education*, Prom. – SG, No. 95 of 08.12.2015, in force as of 08.12.2015. [In Bulgarian: НАРЕДБА № 5 от 30.11.2015 г. за общообразователната подготовка, Обн. – ДВ, бр. 95 от 08.12.2015 г., в сила от 08.12.2015 г.]. Government of Bulgaria, Sofia.

Peev, D. (Ed.) (2015). *Red Data Book of the Republic of Bulgaria. Vol. 1: Plants and Fungi.* Bulgarian Academy of Sciences and Ministry of Environment and Water of Bulgaria, Sofia. http://e-ecodb.bas.bg/rdb/bg/vol1.

Pre-school and School Education Act (in force since 01.08.2016). [In Bulgarian: Закон за предучилищното и училищното образование (в сила от 01.08.2016 г.)]. Government of Bulgaria, Sofia.

Roa, G.R., Krishnakumar, G., Sumesh, N., Dudani, M.D., Subash C. and Ramachandra, T.V. (2014). Diversity and regeneration aspects of medicinal plants at Devimane, Uttara Kannada district, Karnataka, Central Western Ghats. *J. Biodivers. Manage* Foresty, 3(1): 1–8.

Rozensweig, M.L. (1995). *Species Diversity in Space and Time.* Cambridge University Press, Cambridge, UK.

Tsanova, N. (2007). *Standards and Curricula in Biology – Way of Use.* Pensoft Publishers, Sofia.

Yotovska, K., Asenova, A. and Bancheva, H. (2015). Framework to ensure and maintain the quality of non-formal education for sustainable development. *Seminar of Ecology with International Participation, IBER, BAS*, pp. 212–218.

Yotovska, K., Assenova, A. and Kostadinova, H. (2016). *Pedagogical design of the learning process in the electronic environment (on the example of Biology and Health Education, 9th grade, Ecology section)*, 70 pp.

Part III
Environmental Sustainability Education in the Formal Curriculum

Chapter 10
The Development of Environmental Sustainability Education Within Greenland's Higher Education

Gisele M. Arruda

1 Introduction

The end of the Cold War has triggered a new development trajectory for the Arctic region. This has become less militarized and emerged as a new open space for entrepreneurship and economic development; a process stimulated by political support for industrialization based on the abundance of local commodities and their attractive prices. These factors triggered a still not completely regulated expansion of industrial and maritime activities in the region.

The Arctic is the new frontier, a developing region in need of emergent and expanded understanding about its different complexities and dynamics. It has become an extreme case study for climate change, socio-environmental adaptation and education. The search for natural and energy resources, the dynamism of entrepreneurship, new maritime routes for international trade and new extractive industrial activities in unexplored areas are drivers triggering significant societal and developmental transformation. The challenge for education in this region is to understand these drivers of change and respond appropriately to them.

This chapter is the outcome of 5 years of fieldwork in Greenland, which involved visiting urban and remote communities. The aim was to understand the Greenlandic development process and its drivers of change, not least in the face of climate change impacts, and how this dynamic is addressed by Greenland's Higher Education system. The key questions are to what extent is sustainability present in Greenlandic Higher Education governance and curriculum and how this system relates to its contexts – the changes caused by climate change, modernity, autonomy, education and self-determination.

The development of the Greenlandic education system took place in two distinct historical phases: pre- and post-colonization. Before colonization, education was an

G. M. Arruda (✉)
Global Energy and Sustainability Programme, Coventry University, Coventry, UK
e-mail: garruda@anvivo.org

E. Pénzesné Kónya et al. (eds.), *Environmental Sustainability Education for a Changing World*, https://doi.org/10.1007/978-3-030-66384-1_10

informal social activity based on traditional cultural practices, gender-oriented activities with foundations on livelihood and traditional economy and systems (hunting, fishing, crafts, etc) (Aslaksen et al. 2009). After colonization, however, there was a rapid transition from a subsistence-based (hunting and fishing) society with no formal educational system to a complex modern society with new economic, industrial and educational demands.

The structural transformation of Greenlandic society has always involved important occupational and gender transitions leading to economic and educational adaptations. The 1970s were years of intense educational adaptation and prioritization with the semi-autonomous administrative institutions and frameworks introduced by the Danish government. Since the introduction of the Home Rule, primary, secondary and higher education has undergone profound changes in terms of learning goals, language and curricula always oriented by these industrial, occupational and gender transitions. There has always been a concern to promote compatibility from a Greenlandic perspective between the education system (content and frameworks) and the needs of a modern and globalized society. Despite many challenges and setbacks, the goal of the Home Rule government has remained to raise Greenland's education to Nordic levels (Greenland Home Rule 2000). However, there was no reference to Earth systems, climate change, Sustainable Development or Education for Sustainable Development in the mainstream Greenlandic HE curriculum until 2015. Monitoring programmes and scientific data had long been carried out by Nordic scientists. However, after 2015, expertise clusters (expert institutes at the University of Greenland) started to coordinate studies and applied this learning into teaching and learning in a more systematic way. In 2019, the Greenlandic HE system, while continues to be under a transition process, it has already incorporated important sustainability components in order to address historical drivers of change that are transforming Greenland's bio-geo-physical, economic and socio-environmental contexts.

This research considers the methodological challenges of assessing the diverse responses of the Greenlandic population to socio-environmental change, especially given differences in these changes in different areas. Initially, a comparative study of the climatic and socio-environmental impacts was constructed from the formal published literature combined with the grey-literature of regional reports, international conference discussions and 'scientific petit comité'. Insights from this were analysed through interviews with scholars, politicians and local Higher Education authorities' representatives from the Greenlandic government. These data helped create a platform of ideas about the impacts of development on Greenlandic society and how this society creates systematic and contextual knowledge and response to the different drivers of change.

2 First Driver of Change: Climate Change – The Complexity of Contextual Scientific Knowledge and Assessment of Greenlandic Cryosphere (Greenland Ice Sheet Decline)

Greenland, as the largest island in the world covered by inland ice, has a key role in determining the climate in the Northern Hemisphere. Its geographical features and the presence of important glaciers like Kangerlussuaq Glacier (the largest glacier on the east coast of the Greenland ice sheet) and Jakobshavn Glacier (located in the west coast near the Greenlandic town of Ilulissat) make it a perfect location for monitoring patterns of change in the cryosphere and ice sheet. Greenland's position in the North Atlantic, within an ice-filled ocean, its elevation, and the sea currents from the Arctic Ocean create specific climatic dynamics for the island and the surroundings which have been monitored in land by meteorological ice stations and via satellite remote sensing. Programmes like Copernico, CryoSat-2, GRACE-FO, EOS AM-1 satellite, operated by NASA, with the MODIS and ASTER sensors are fundamental to understand changing process on regional and global scales (PROMICE 2019). These tools assist researchers in their interpretation and analysis of the scale of glacier velocities, glacier retreat, ice sheet loss and broader climate change programmes. Due to this technology, it was shown that Greenland's fastest glacier is the Jakobshavn Glacier, which drains approximately 6.5% of the Greenland Ice Sheet (Fig. 10.1) (Joughin et al. 2004; Fausto et al. 2016). However, despite these modern technological tools for investigation, data collection and interpretation of the mechanisms that trigger ice loss, the mass balance fluctuations of the Greenlandic Ice Sheet are still insufficiently understood, although the trend indicates the acceleration of ice sheet decline. Ice sheet and glacier behaviour in Greenland is a sensitive indicator of climate change and it is specially significant to infer quantitative patterns of change in other domains like global temperature, jet-stream, rising sea levels and declines in global fresh water (Oltmanns et al. 2018).

Decline in the Greenlandic Ice Sheet has been studied intensively since 1840. Figure 10.2 displays the current pattern and compares the ice sheet's surface gains and losses in the period 1981–2010 with those recorded in September 2018 until 23rd February 2019 (Box et al. 2009; Mankoff et al. 2018). This estimate follows the Regional Climate Model HIRHAM5 (Langen et al. 2017) and does not include the mass lost when glaciers calve off icebergs and melt as they encounter warm seawater (Bevan et al. 2019).

These patterns observed in the glaciers velocity, the accumulation zone of the Greenland Ice Sheet and the fluctuation of the surface mass balance describe live and dynamic natural systems and represent important sources of climate change data to be interpreted by the international academic community. This kind of research as well as its implications for climate change has been conducted, primarily, by foreign Earth Systems and environmental scientists from Europe and US. However, the information also has great significance for local academic investigators who aspire to orient local responses to this important driver of change and understand its impacts on other local systems (Fig. 10.3).

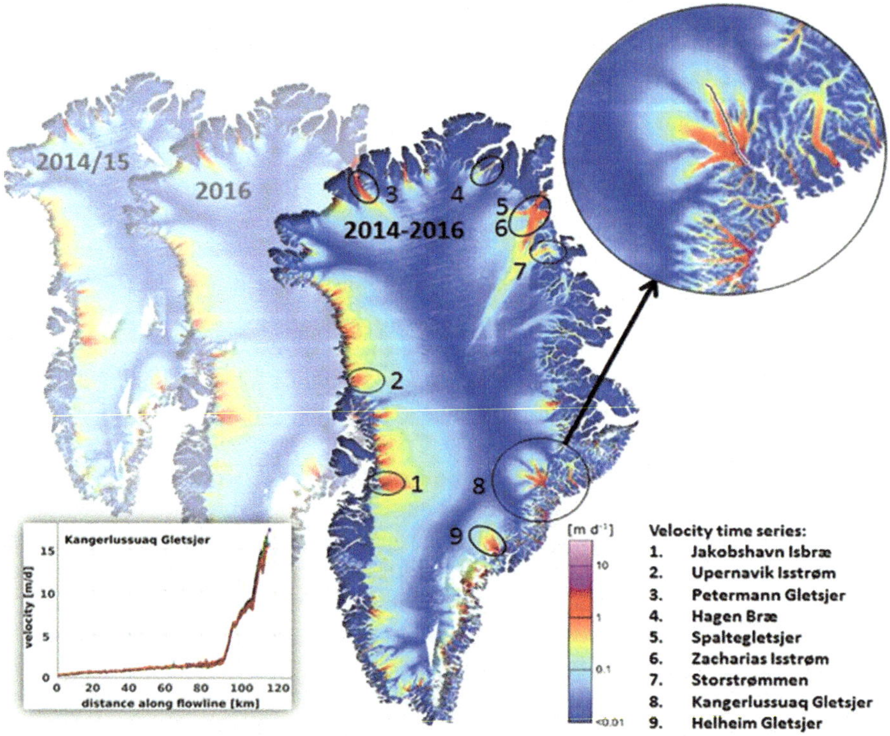

Fig. 10.1 Greenland Ice-cap Velocity time series. Source: ESA CCI (2019)

The cryosphere is an element of the Earth System consisting of water in its seasonally and perennially frozen state (Hovelsrud et al. 2011, p. 2). Its components include snow, solid precipitation, permafrost areas, river and lake ice, glaciers and ice caps, the Greenland Ice Sheet and sea ice (Cogley et al. 2011; Cuffey and Paterson 2010). These cryospheric components form a unique Greenlandic system that is intrinsically interconnected with the ecosystems, social systems and global climate. Changes affecting the Arctic cryosphere alter, significantly, the functioning of local, regional and global socio-environmental and economic systems. Hence, in the last 5 years, Greenlandic Higher Education has started to develop programmes of study and research and to produce scientists with expertise in climate change, Earth sciences, and the human dimension of change. This sets high importance on developing a transdisciplinary HE curriculum that comprehends the bio-geo-physical, economic and socio-environmental interconnections and that informs local communities.

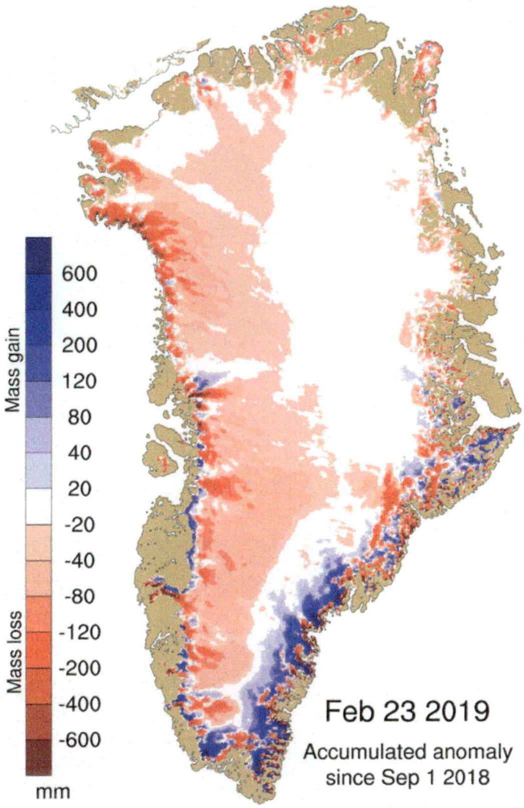

Fig. 10.2 Accumulated anomaly of Greenland's Ice Sheet mass loss according to Regional Climate Model HIRHAM5. (Langen et al. 2017)
Source: http://polarportal.dk/en/greenland/surface-conditions/

3 Second Driver of Change: Greenlandic Context of Modernization

The Second World War was a major turning point in modern Arctic history, not least because of the separation of Greenland, Iceland and Faroe Islands from Denmark. It was also marked by so-called 'friendly invasions' of allied forces, especially in Greenland, which hosted an American strategic air base in the early years of the war, whose powerful indirect influence instigated nationalist sentiments as well as a new national consciousness. This period marks the beginning of 'Home Rule' (Direktoratet for Kultur, Uddannelse, Forskning og Kirke 2004) implying local decisions on cultural, social, health, environmental and economic issues, but, tacitly, the continuation of the dominance exerted by Denmark. Indeed, from the 1721 colonization of Greenland by the Denmark-Norway coalition to the 1950s trading activities of Copenhagen's Royal Greenland Trading Company, ideas from the outside world have driven profound societal change in indigenous Eskimo culture, social order and institutions. According to Jennes (1968, p. 47):

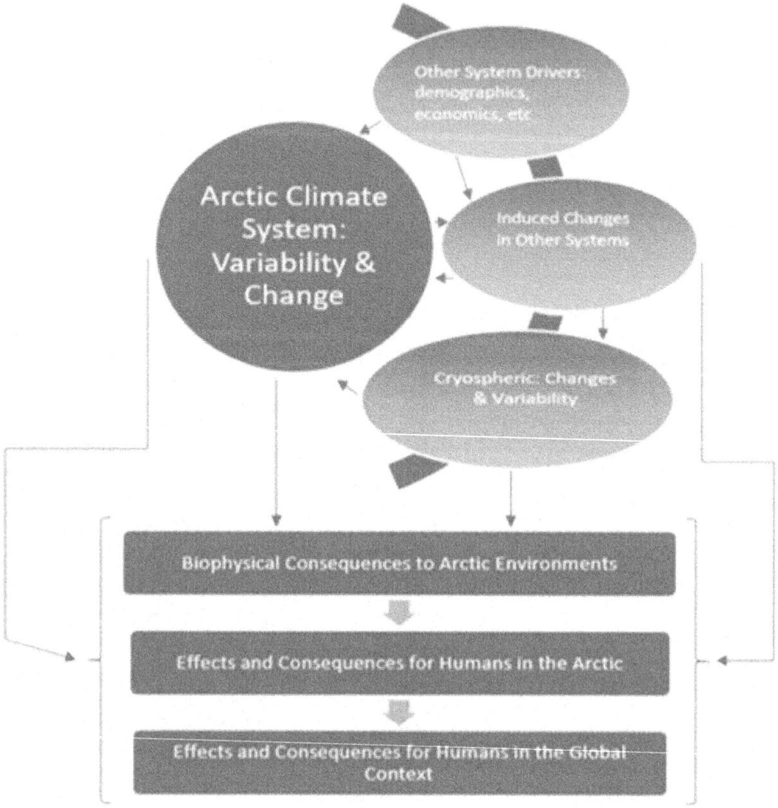

Fig. 10.3 The Arctic cryosphere and linkages to various levels of effects or consequences of changes. *Source:* Hovelsrud et al. (2011)

> in no other region of the Arctic have Europeans deliberately trained Eskimos to lead their people up civilization's steep path [...] only in Greenland have the educational authorities [...] deliberately trained Eskimos and their descendants to lead their people; only there the Eskimo population has been successfully integrated into today's world; a world of modernity, democratic reforms, and a new entrepreneurial spirit offering all the nice goods, which the rest of the world had been using for more than 200 years.

These developments have also created most of the social problems that currently afflict modern Greenlandic society. Greenland is a post-colonial society that has undergone a very rapid modernization process. This has brought benefits that include advances in infrastructure, health care, education and welfare at the cost of the dis-benefits typical of such accelerated development: alienation, urbanisation, and capitalisation (Lee and Newby 1983).

This situation has two particularly important aspects: its impact of development on Greenlandic society and the ways this society has responded to change. Equally fundamental is understanding how these responses align to sustainability principles and how these responses enable adaptation and social equality. Greenland's recent history raises several ethical dilemmas. Modernity contains an intrinsic paradox of creation and destruction, especially of traditional cultural frameworks, which can present societies with serious dilemmas and, sometimes, an existential crisis. It may bring on board powerful technologies to promote well-being but, at the same time, it can provoke profound crises of reality, truth and identity in multi-cultural societies (Arruda 2018b, p. 107).

Most of the economic activities being developed in Greenlandic communities show the traditional pattern of male-dominance. Industries like hydrocarbon (oil and gas) exploitation, mining and infrastructure generate (currently) 77% of male employment opportunities and thereby reallocate, to the new economic hubs, a great parcel of the social capital previously dedicated to traditional activities (Kleist et al. 2015; Government of Greenland 2009). On the other hand in Higher Education, since 1990, the number of women has exceeded the number of men completing a degree and those who have studied beyond the level of a baccalaureate. Courses like journalism, social work, teachers' training and nursing are dominated by women while vocational education and training are still dominated by men (Poppel and Chemnitz 2009). This context has created severe imbalance in terms of traditional family structure, gender and occupational roles and the proportion of higher educated citizens.

In recent years, the discovery of vast energy and mineral resources in previously inaccessible iced-covered areas of Greenland have sparked much debate about the risks and benefits of extractive industry. Local communities in these resource-rich areas are increasingly exposed to the external pressures of development advocated by governments and their industry partners. Modern infrastructural development is part and parcel of the developments that facilitate resource exploitation. New ports and new maritime routes across the Arctic Sea are being made possible by the decline in sea-ice. However, even in the more developed urban centres of Greenland, there is growing concern about the real costs of becoming an extractive mining economy and recognition that this requires a balanced rethinking of sustainability issues, including their economic, environmental and socio-cultural pillars (Elkington 1994).

Despite the strategic environmental and social risk assessments (SEA/EIA and SSIAs) prepared by energy and mining companies towards gaining licenses to exploit resource-rich areas (Arruda and Krutkowski 2016, 2017) and, despite the specific regulations regarding transboundary pollution and environmental toxicity according to the best international standards and practices (NORSOK – Norwegian safety Regulations; Environmental Benefit Analysis) (NORSOK 2010), there is no systematic Greenlandic sustainability policy in place to guide industries or planners. Companies apply their own strategies of environmental management and infrastructure development and civil society works to promote greater understanding of the impacts of large-scale resource projects through research, information sharing and public consultation but not framed by a comprehensive and systematic governmental sustainability policy. Companies apply their own strategies of environmental management and infrastructure development, while, in the absence of a comprehensive and systematic governmental sustainability policy, civil society is left to promote greater understanding of the impacts of these large-scale resource projects through research, information sharing and public consultation. Inevitably, this is more than civil society is capable of mitigating and this is reflected in the practical socio-environmental consequences for Greenlandic society. Of course, ideally the Greenlandic government in partnership with its main trading nations (Denmark, EU, US, China), who aim to benefit from exploiting Greenland's natural resources, should work together to ensure the delivery of the United Nation's framework of Sustainable Development Goals (SDGs) (United Nations General Assembly 2015). However, it is not clear that this is happening effectively.

In theory, Sustainable Development means human development. It involves the adoption of appropriate standards oriented to those local values that play a fundamental role in the process of promoting well-being in the community (Larsen and Fondahl 2015; Poppel 2006, 2015). Often, the Western concept of development seems to conflict with local lifeways. Greenlandic communities do not recognise their portrayal through the non-native lens that is applied to them, despite the lip-service to inclusivity of the enterprises (OECD 2011) promoting new technology, the trading of manufactured goods and infrastructural projects. Today's Greenlandic development is conflicted by competing values and cultures. The result is a process that is leading in the opposite direction to freedom of choice and Sustainable Development, at least in the terms defined by the World Commission on Environment and Development (WCED 1987). This can be evidenced by Figs. 10.4, 10.5 and 10.6, which illustrate the absence of a systematic sustainability policy, infrastructure and of a sustainable mindset towards the environment.

Sustainable Development has an interdependent relationship with human development, and it is promoted through the adoption of appropriate standards oriented to values that play a fundamental role in the processes of deliberating on well-being (Larsen and Fondahl 2015; Poppel 2015). Currently, Greenland's development is focused on the life cycle of extractive projects. It is not oriented towards a systematic, holistic and diversified model of development framed by the UN's Sustainable Development Goals (SDGs). However, there may be signs of positive change.

Fig. 10.4 Ilulissat town centre. (Picture: Gisele M. Arruda)

Fig. 10.5 Ilulissat town centre and the market area. (Picture: Gisele M. Arruda)

Fig. 10.6 Ilulissat town centre. (Picture: Gisele M. Arruda)

Recently, Greenland has undertaken an innovative approach to reduce its reliance on electricity produced by diesel fuel by developing the Ilulissat Hydroelectric Project. This is a mega hydropower project driven by glacial meltwater and the discharge from natural glacial lakes. Here, meltwater is channelled 200 m through the permafrost to underground turbines in a sea-level powerhouse (Arruda 2018).

4 Third Driver of Change: Greenland's Autonomy and Education

In international law, Greenland is a self-governing Arctic country (United Nations General Assembly Resolution 1962), which nevertheless remains, formally, part of the Kingdom of Denmark. Greenland's education remains intrinsically linked to the Danish educational system despite Greenland having taken responsibility for its own educational governance since the 1980s under the Home Rule system (Gulløv 1979). The mantra of Greenlandic local policy is 'autonomy' and 'independency', a mind-set revealed by several interview reports. Greenland, seeing itself as a post-colonial nation since the 2009 Self-Rule Act (Government of Greenland 2009), is actively seeking control of its development opportunities and its right to manage natural resources. These issues are clearly reflected in Greenland's Higher Education system. Interviewees are under code names to ensure confidentiality as per Table 10.1.

Much is inherited from Greenland's colonial history and a colonial system based on ethnic and social stratification (Sørensen 2006). Some is the product of Christian missionary works, which supported education 'and protection' by the 'mother land' (Marquardt 2010). This social stratification created inequality (Greenland Home Rule 2003), which still persists, as Danish people dominate the government as high income civil servants with mixed-blood Greenlanders in lower status, lower income appointments.

With increasing autonomy and independence, new educational policies and higher standards of local education were required to enhance the number of graduates produced by Greenlandic Higher Education. The European Commission (2013) recommended that output needed to reach 100 graduates per year to avoid a potential shortfall of 2000 graduates by 2025 (EU Commission 2013; Fig. 10.7). This

Table 10.1 Interviewees and quotes

Interviewees	Quotes
Dina (Inerisaavik – Pedagogical support staff)	'The curriculum was designed in the 1970s and it has been reviewed since then' (...) the curriculum has no evident links with environmental sciences. It has been a long period of reform but since May 2015 significant changes started to happen and I need to figure out the basis, motivation and outcomes of it.'
Patrik (education Ministry's advisor)	'The official report from the Greenland's Ministry of Education, church, culture and gender equality (the education plan II) created reaction in relation to the whole educational systems in terms of leadership, teaching and parenting.'
Birgit (higher education expert in pedagogies and curriculum)	'Environmental sciences are part only of the curriculum for children and youth but not part of the higher education curriculum yet.'
Erna (educator)	'The reform also focuses on the Personal Development Plan (PDP) to tackle social, psychological and internationalization issues.'

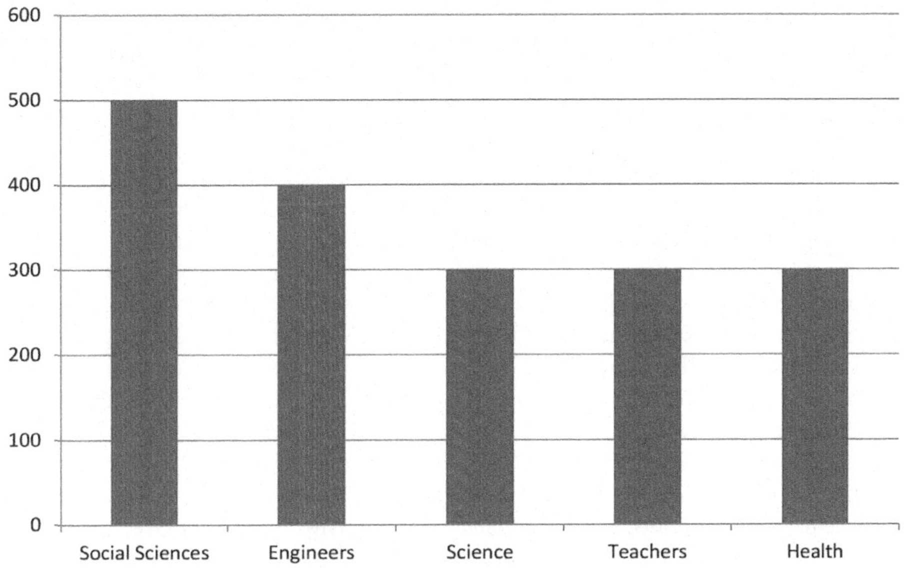

Fig. 10.7 Expected shortage of graduates in 2015 for the main domains of Higher Education. (Adapted from EU Commission 2013)

challenge continues and there is a lack of professionals in almost all subject domains (Elmore 2006; Danish Agency for Higher Education 2014). The problem is compounded by low application rates from students graduating from secondary education, who do not feel it worthwhile to continue their studies. Part of the problem is that, in practice, students entering Higher Education still need to be proficient in both Danish and English.

The contradictions generated by the intermingling of Greenlandic and Danish cultures make it possible to argue that 'modern Greenland is understood as being "Danish" rather than "Greenlandic" because 'Greenlandicness' (Thomsen 1998) often refers to 'old' traditions of hunting, fishing, story-telling, etc., while modernity is considered "Danish" (Brincker 2017; Thomsen 1996). In words of Bjørst (2008, p. 29) 'it may seem difficult to be both Greenlandic and modern' at the same time. This dichotomy is the source of a major mismatch in Greenland's Higher Education with effects on other areas of people's lives because for a young person living in Greenland there is a huge distance between the everyday life at an education institution and his or her family life of ancestral routes. The traditional and the modern, living side by side, an educational mismatch with several social implications.

The sense of autonomy and self-determination also accelerated educational policy making and intensified the need for educational outcomes to address the expected shortage of graduates. This has created an extra layer of pressure on matters relating to identity, autonomy and self-determination. Local documents, interview reports and local literature all argue that the curriculum of the whole education system in Greenland is a product of the 1970s. Subsequently, the curriculum is being subject to government review in an attempt to build capacity and support national educational updating and achieve reform that is sensitive to Greenland's geographical complexities, where most settlements are very small, isolated, and very different to towns like the capital Nuuk, whose population of circa: 17.984 represents almost a third of Greenland's total (City Population 2019).

Greenland's first significant curriculum reforms began in 2002 and developed continuously until the present (Wyatt 2012). The main concern has been to close the cultural gap in an educational system that contributes to young students' reluctance to enter Higher Education. This challenge calls attention to the importance of cultural compatibility in indigenous education (Demmert and Towner 2003; Rivera and Tharp 2006) and it was the base for a nation-wide reform aimed at creating culturally compatible education. This work was based on the Standards for Effective Pedagogy developed at the Center for Research on Education, Diversity, and Excellence in Hawaii (CREDE 2019). The CREDE standards (Table 10.2) represent principles of effective teaching and learning derived from studies conducted in other indigenous communities that aim to promote 'cultural congruity' (Wyatt 2010). These have been applied to help enhance the linguistic abilities of Greenlandic students' and support their native cultural identity with the purpose of improving their employability and Greenland's labour market.

In recent decades, most educational initiatives have been funded by the EU and by Denmark due to its close relationship with Greenland, which provided the necessary funding for reform, but, later, this relationship was reviewed, renegotiated and the level of support reduced. With less available funding and in the face of economic instability, inflation, and an ageing population, many people started seeing the new energy and mineral developments as a unique opportunity to be independent from Denmark and stabilize the local economy.

The year 2015 became a second landmark date for educational reform in Greenland and educational development became a political priority, according to the Ministry of Education secretariat. The Ministry of Education focused its new strategy on the integration of curriculum of the whole educational system, not only Higher Education but also, on preschool, primary and secondary, high-school, education targeting children's and youth education, in part, to increase progression to Higher Education. The belief remains that investing in the culturally compatible education of children and youth will increase entry to Higher Education (Wyatt

Table 10.2 CREDE Standards for teaching to promote cultural congruity

(a) Joint productive activity	Teachers and students collaborating on joint products
(b) Language and literacy development	Teachers supporting language development in all classrooms and subjects
(c) Contextualization	Teachers making connections between students' prior knowledge and new information
(d) Complex thinking	Teachers supporting students' engagement and skills in critical thinking
(e) Instructional conversation	Teachers instructing through dialogue

Adapted from CREDE (2019)

2009; Kahlig and Banerjee 2007). However, progress continues to be constrained by inefficient implementation.

One key focus is language learning. Just 10% of the Greenlanders consider themselves Danish while 90% see themselves as Greenlanders, typically a mix of Inuit and Danish. Basically, 90% of the Greenlanders are mixed (Kleemann 2018) and, for this reason, the expression 'indigenous people' is not employed in Greenland because most Greenlanders have Inuit and Danish ancestors. However, as interview findings confirm, there is one school system for all.

The government's motivations for driving educational reform are economic development and autonomy from Denmark. For example, from 2002, the curriculum subject called 'Christianity' was replaced by one called 'Religion & Philosophies'; the reason was to remove a term associated with the missionaries of Greenland's colonial period. However, language instruction was a key target of the larger reform process. Presently, the language of instruction is Greenlandic, while Danish and English are used in stages 8, 9 and 10. The reforms also introduced Personal Development Planning (PDP), in order to reorient learners to their future potential in the modern economy as well as to help them tackle social and psychological issues of a complex society based on multicultural asymmetries and educational mismatches.

Since the 18th of May 2015, the Greenlandic educational system has experienced a real transition because its most important educational department at University of Greenland's 'Institute of Learning' became a governmental entity (TMECCGE 2015). This institute is responsible for the education and training of new teachers/lecturers as well as for curriculum design and review at all levels. The Institute of Learning is part of the Danish government hierarchy and tasked with implementing a curriculum reform based on an important official report from the Greenland's Ministry of Education, Church, Culture and Gender Equality entitled 'The Education Plan II' (TMECCGE 2014a, b), which aimed at evaluating the Greenlandic curriculum and education system.

One of the resulting reform's main goals was to create a more flexible educational system in which high schools offered a common core curriculum with the opportunity to choose supplementary subjects based on students' individual interests and skills, in order to promote progression and retention towards tertiary education. In the Greenlandic government's view, it was important to educate the youth to prepare them for the labour market and to be active participants in Higher Education. Moreover, the government had in mind to establish more education programmes (professional bachelor's degrees) based on a more internationalized range of disciplines involving human rights, diversity and identity, environment and climate change. Until May 2015, the HE curriculum had been unchanged since the 1980s. Its core themes related to Greenlandic grammar, Greenlandic literature, Greenlandic history and political science within a Greenlandic framework, local social and gender issues but it ignored environmental sciences and its cognate disciplines. An important feature of Greenlandic HE has been an imbalance between a predominance of students in social sciences and small minority in technical and scientific disciplines. For this reason, the Government reforms also encouraged expansion of

Greenland Institute of Natural Resources at the University of Greenland. The aim was to enhance capabilities and reform curricula to provide students with knowledge of mechanisms and processes affecting their lives and livelihoods, such as rights to land, natural resources, economic activities and impacts of climate change. These were proposals that triggered the education reform process after May 2015.

An additional goal of this reform process was to reduce the cultural asymmetries and tension in the educational system by creating a culturally compatible educational environment that strengthens native Greenlandic culture and identity. This educational reform has proved more difficult to implement because, at ministerial level, the greater concern is for economic development and its stimulation.

In May 2017, it was possible to see the first concrete signs of a broader national discussion on Sustainable Development. This was materialised by Denmark's signature as a participant in the 'Fairbanks Declaration 2017' by signing the 'Agreement on Enhancing International Arctic Scientific Cooperation' (Arctic Council 2017). This was a true landmark in initiating a debate on Sustainable Development and on the United Nations' Sustainable Development Goals (SDGs) in Greenland with sight on improving economic and living conditions, efforts to reduce transboundary pollution (POPs) (de Wit et al. 2019) and black carbon, the value of sustained biodiversity monitoring (Arctic Biodiversity Assessment) and traditional knowledge, innovative energy and ICT infrastructure (internet), a research-focused curricula, data sharing on climate change indicators, and of course, the special focus on local updated assessment of snow, water, ice and permafrost (SWIPA 2008). These are guidelines that represent a shift in Greenland's perspectives on educational reform that impact not only the curriculum but also Greenlandic teacher training and qualification at a post-graduate level.

5 Greenlandic Higher Education and the Development of Sustainability Education

Greenland has experienced a process of evolution comprising a shift from a traditional subsistence economy towards a more market-oriented. Educational reform has similarly been linked to supplying the labour market and the promotion of trade and industry (EU Commission 2014).

The most important Higher Education institution in Greenland is the University of Greenland or *Ilisimatusarfik*, which was established in 1983 in the capital Nuuk and granted University status in 1989. The University's early activities involved delivering teaching and research in literature, languages, administration and theology (EU Commission 2014). A new board of studies was created by the Greenlandic Parliament in 2008 and, currently, the institution offers 11 university undergraduate degrees with individual courses in Danish, some in Greenlandic and a very few in English. In 2018, it enrolled 205 students and around 14 teachers, a small number, but many Greenlandic students choose to attend university abroad, especially in

Denmark. Since 2015, it has also hosted an international and interdisciplinary master's programme on 'West Nordic Studies, Governance and Sustainable Management', which has been developed in partnership with four official international partners. These are the University of the Faroe Islands, Nord University, University of Iceland and the University of Akureyri in northern Iceland.

Another key player in Greenland's Higher Education is the University of the Arctic (UArctic), a cooperative network of educational colleges and institutes, which aims to promote northern voices in the globalizing world and promote partnership between native and non-native inhabitants. UArctic's academic supporters include 143 member institutions, not least Dartmouth University, USA, University of Lapland, University of Oulu, Finland, and Memorial University, Newfoundland. UArctic has three members in Greenland. These are *Ilisimatusarfik* (University of Greenland), the Greenland Institute of Natural Resources, also located in Nuuk, a research institute established in 1995 that focuses on marine resources and climate change, and *Perorsaanermik Ilinniarfik* (College of Social Education), which is located in Ilulissat and teaches education, including preschool, child and youth care, social problems and social welfare. In 2017, the college co-hosted a conference on 'Reconciliation, Learning and Cultural Diversities' which was co-hosted by the Greenland Reconciliation Commission and that focussed on reconciliation and decolonisation.

Also relevant is the Arctic Technology Centre of the Technical University of Denmark (ARTEK), which is based in Sisimiut, Greenland's second town of some 5500 people located around 200 miles north of Nuuk. Since 2000, ARTEK has offered undergraduate courses in engineering and Arctic technology, Fishery Technology, as well as a Masters in Cold Climate Engineering. The Centre is a joint venture with *Teknikimik Ilinniarfik*, KTI (Technical College of Greenland), and aspires to recruit 200 students. Its courses in Arctic technology address important challenges related to the need of sustainable infrastructure and it hosted the conference on "Urbanisation and infrastructure in the Arctic – challenges to sustainability" in 2014 (ARTEK 2014), which dealt with 'the fundamental challenges of planning and building in a cold climate'. Typically, the languages of instruction are primarily Danish and Greenlandic, but again, since 2015, some courses are available in English.

During interviews, Greenlander respondents report that they see the curriculum, from pre-school to University as a whole, one body of components. Respondents also called attention to the curriculum's continuing lack of major alignment to the environmental sciences, except in very specific cases. Some natural science is taught, for example, in Sisimiut, where the engineering course includes elements of environmental science. However, more generally, the mainstream Higher Education curriculum still does not deal with either climate or wider environmental change or energy studies. Both documentary and interview evidence, obtained from the Institute of Learning in Greenland University, confirm that Environmental Education and ESD are not yet part of the Higher Education curriculum.

Environmental education does seem to be embedded – to a certain extent - in the Secondary and Primary curriculum. However, until recently, the Higher Education curriculum had been unchanged from the 1970s and 1980s and preserved the concerns of those decades. In May 2015, significant changes were proposed to implement a greater focus on indigenous pedagogies, contextualization, cultural activities important for enhancing students learning and engagement (e.g. Inuit dance, art, music and craft), but these components are still insufficient in this curriculum. Indeed, the reforms of May 2015 proposed a more focused consideration of the ways that students learn, how to enhance learning and how to implement life-long learning in the Greenland context. Recently, these have started to be implemented, along with proposals to develop educators' training on how to teach multidisciplinary subjects (involving technical, scientific and sustainability know-how), an approach that has received more support from the Ministry of Education in concerted initiatives with international partners.

Most interviewees report that the May 2015 educational legislation is well-structured, but has just not been completely implemented because it is an on-process consisting in assessing gaps, accommodating needs and styles of learning and embedding sustainability. They feel that effecting its implementation to the point of being practiced will still take more than 20 years because the tutors need to be re-educated and re-trained, a notion which is resisted by many current tutors.

Throughout, interviews reveal a clear mismatch between the colonial and post-colonial mind-sets, which can only be addressed by adopting more culturally inclusive models of education. One possibility, that in one interviewee's opinion rendered positive results, is based on Vygotsky's socio-cultural theory (Vygotsky 1978; Fig. 10.8), and would include cultural elements of contextualization, personal development, personal worldview (family experience and roots), and collectivist communication (body language).

Documents and interview findings collected from the Institute of Learning include no reference to the word 'sustainability' in the curriculum (*laereplan*) or in the education system until levels 6 and 7 for 'historical reasons'. However, the term 'sustainability' does appear in a recently designed Higher Education curriculum on 'Environmental Change' but this is not yet embedded in the curriculum. This curriculum also emphasises life-long learning, which, despite being part of the legislation, is another concept in need of further development in Greenland.

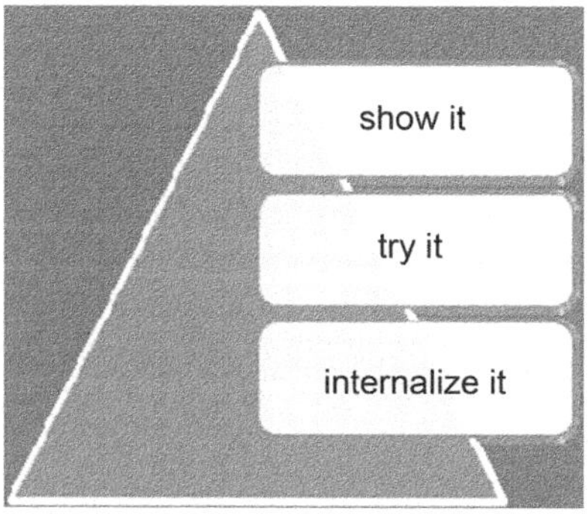

Fig. 10.8 Inclusive model of education
Source: Spindler (1997)

	2003	2004	2005	2006	2007	2008	2009	2010	2011	2012	2013	2014	2015	2016	2017	2018
Upper secondary education	281	288	347	330	356	457	544	480	506	542	551	611	576	510	490	500
Vocational education and training	703	642	566	689	658	719	722	751	842	795	777	802	802	890	836	811
Supplementary examination courses	46	49	57	64	63	67	81	85	115	93	121	132	125	114	113	103
Short-cycle higher education	50	49	75	58	63	82	79	82	73	86	85	114	107	126	123	145
Bachelors programme	87	78	89	87	99	95	115	133	120	141	149	133	149	136	123	125
Professional bachelors programme	133	143	138	158	145	185	158	217	267	258	215	278	263	212	165	222
Masters programme	15	20	24	27	38	22	33	44	35	36	47	45	54	49	57	42

Fig. 10.9 Enrolment in Greenlandic education system from 2003 to 2018
Source: Statistics Greenland (2019)

In terms of educational strategy, interviewees expressed the opinion that the government and the departments engaged in educational reform are on the right track. The number of students attending Higher Education has grown considerably and statistical data shows that a large percentage of the enrolment in vocational education supports emergent industries (Fig. 10.9). Statistics are positive in terms of the number of students that entered the educational system, the problem is to know how many will complete it and how many will adapt to the opportunities and challenges presented by Greenland's rapidly changing economy and society.

The reform carried out in 2015 promoted significant changes in the Higher Education structure and introduced important discussion on climate change, energy and environment in the context of sustainability. Interviewees believe that with the implementation of the new 'Environmental Changes' curriculum and through the newly launched degree programme on 'Society', these topics and their interfaces may become more comprehensively addressed in Higher Education, thanks in part to the influence of the United Nations' SDGs in 2017 and the signature of Fairbanks Declaration.

Meanwhile, Greenland Ice Sheet predictions combined with specific cases of pollution and the visible effects of climate change have initiated an intense local debate on how to teach people, both locally and around the world, how to deal with climate change and other environmental challenges. A pilot study was conducted recently to examine educators' perception about climate change and environmental issues in Greenland because, as it was reported, there is a misconception about these topics in the local media. This suggested that Greenland's educators may be concerned about climate change, but this is less true of the common citizen, as the volume and types of waste seen in the streets of Nuuk and Ilulissat demonstrate. To effect change, educators need to be enabled to teach about the subject in theory and to be engaged with practice. Fieldwork reveals that, while there is a need for teaching the theory and practice of climate change, environment and sustainability, there are obstacles in the way, which include language issues and the lack of suitable professional training for teachers. This study included data from a roleplay among a group of four educators on the subject of hydropower and how it works in Greenland. Basically, the roleplay revealed that among the few participants that attended the session, one didn't understand, one understood but couldn't explain the theory, one understood but was not able use the information in practice, while just one could explain and apply the material. The study demonstrated the challenges of teaching and training Greenland's educators and students in the absence of contextualization and a holistic view on local drivers of change. It was concluded that the real problem (in terms of the number of students in HE) was not the students but their tutors, who were trained long ago in the Danish system and hadn't adapted to the new requirements of a changing educational system and the local drivers of change including clear notions of Sustainable Development.

Sustainable Development and education have an intrinsic, interdependent relationship with each other, largely because no sustainable society can operate without high quality education as well as human development being intrinsically linked to education standards (AHDR 2004, 2015; Poppel 2015). Curricula which only focus on key competencies for the world of paid employment is deficient (Marsh 2009) because learners are persons not only industry professionals. ESD is a vision of education that seeks to balance human and economic well-being with cultural traditions and respect for the earth's natural resources. ESD involves "methods and approaches to develop lifelong learning; fosters respect for human needs that are compatible with sustainable use of natural resources and the needs of the planet; and nurtures a sense of global solidarity" (UNESCO 2005, p. 2; Ryan 2011, p. 3).

The positive aspect is the fact that these challenges have triggered an articulated process of cooperation among national and international stakeholders (scientific, academic, corporative) to allow for the establishment of scientific study programmes and the expansion of physical capabilities as laboratories, research institutes, business incubators that have been strengthening multidisciplinary research within society, environment and health as well as giving advice on the sustainable use of natural resources.

6 Conclusion

The local and global implications of the melting of Greenland Ice Sheet and the benefits and costs of a complex and changing Greenland with the multitude of interfaces and impacts on different local and global systems represent an important stimulus to advance Greenlandic Higher Education. A systematic multi-sectorial approach for enhanced research and monitoring efforts that expand understanding of the development process and impacts on Greenlandic ecosystems and society is needed to provide a range of adaptation options and the effective engagement of stakeholders including communities, policy-makers, educators, industries and common citizens. This initiative must go beyond the narrow curriculum versing about environmental management and infrastructure development impacts. It requires a concerted participation of stakeholders, information sharing, and local knowledge co-production framed by an inclusive, comprehensive and systematic sustainability governance and educational policy.

Capacity building and education oriented to sustainability are key components to start designing the future of a more balanced Greenlandic development model. The educational shift in direction promoted by the 2015 educational reform associated to the introduction of the United Nations' Sustainable Development Goals debate in 2017 are the beginning of a process that could be consolidated by the introduction and adoption of Education for Sustainable Development (ESD) parameters.

Higher Education has a pivotal role in promoting the transition from an environmental impacts' perspective – based on the extractive industry mind-set - to a well-structured and articulated sustainability perspective based on the knowledge, dissemination and application of the Sustainable Development Goals (SDGs). It would be through the UNESCO ESD framework that this sustainability perspective can be shaped and practiced in the Greenlandic educational system as a whole. An educational system based on ESD can contribute to analyse the comprehensively complex contexts and drives of change in course through multidisciplinary research, cultural inclusiveness by addressing the urgencies of the present and the needs of the future under a durable and long-term perspective.

Recommendations to enhance Higher Education in Greenland would involve promoting collaboration between Greenland's educational institutions and communities to break historical patterns, de-colonize the curriculum and prepare the nation for an uncertain but more self-created and independent future. Recommendation one would involve stepping further away from long-established, one-sided, colonial education and co-constructing a new educational structure at the interface between Greenlandic modernity and its traditional knowledge systems through culturally compatible education (Wyatt 2009). This would encourage more young people to engage with Higher Education and build the skill-base that Greenland needs (Kahlig and Banerjee 2007). Recommendation two would be to enhance teaching and learning about the interconnections between social sciences and environmental sciences, technical and scientific knowledge creation, to ensure the delivery of an Arctic education that is both relevant and interdisciplinary and that delivers knowledge and competences in the fields of climate change adaptation, polar research and monitoring systems as well as benefits for the new economy.

Finally, it is essential that Greenlandic Higher Education does more to develop a greater understanding of what Sustainable Development means for the wider community of Greenland; how it might support its population, economy and sensitive ecosystems and how it may be possible to promote and achieve the SDGs. It will be necessary to evaluate the educational system as a whole and apply mechanisms to assess levels of adherence to Sustainable Development Literacy (SDL) and to start the process of embedding ESD into the mainstream curriculum by integrating insights from Greenland's specialised institutes of polar research into the content, approaches, curricula and pedagogies of education using the life-long learning educational model. This could be the start of a new era for autonomy, education and self-determination.

References

AHDR-Arctic Human Development Report (2004). Akureyri: Stefansson Arctic Institute.

AHDR-HH (2015) 'Arctic Human Development Report II', Norden, Copenhagen

Arctic Council (2017). Fairbanks Declaration. Arctic Council Secretariat available at https://oaarchive.arctic-council.org/handle/11374/1910 (accessed 10 July 2019).

Arruda, G.M. (2015). Arctic governance regime: the last frontier for hydrocarbons exploitation. *International Journal of Law and Management*, 57(5), 498–521.

Arruda, G.M. (2018). *Renewable Energy for the Arctic: New Perspectives*. Routledge, Abingdon.

Arruda, G.M. (2018b). Artic resource development: A sustainable prosperity project of co-management. *In:* Arruda, G.M. (2018) *Renewable Energy for the Arctic: New Perspectives*. Routledge, Abingdon, p. 112.

Arruda, G.M. and Krutkowski, S. (2016). Arctic governance, indigenous knowledge, science and technology in times of climate change: Self-realization, recognition, representativeness, *Journal of Enterprising Communities: People and Places in the Global Economy*, 11(4), 514–528.

Arruda, G.M. and Krutkowski, S. (2017). Social impacts of climate change and resource development in the Arctic: Implications for Arctic governance, *Journal of Enterprising Communities: People and Places in the Global Economy*, 11(2), 277–288.

ARTEK (2014). *Urbanisation and Infrastructure in the Arctic Challenges to Sustainability. Proceedings of the 10th Artek Event*. Sisimiut, April 2014, Greenland Arctic Technology Centre, DTU Technical University of Denmark, March 2014.

Aslaksen, I., Dallmann, W., Davin, L.H., Høydahl, J.K., Poppel, B., Stappleton, M. and Turi, I. (2009). Interdependency of subsistence and market economies in the Arctic. *In:* Solveig Glomsråd and Iulie Aslaksen (eds.) *The Economy of the North 2008*. Statistisk Sentralbyrå/ Statistics Norway (2009), pp. 75–98.

Bevan, S.L., Luckman, A.J., Benn, D.I., Cowton, T. and Todd, J. (2019). Warming of SE Greenland shelf waters in 2016 primes large glacier for runaway retreat. *The Cryosphere Discuss*. doi:https://doi.org/10.5194/tc-2018-260. Available: https://www.researchgate.net/publication/330092120_Warming_of_SE_Greenland_shelf_waters_in_2016_primes_large_glacier_for_runaway_retreat (accessed29 October 2019).

Bjørst, L.R. (2008). En anden verden: Fordomme og Stereotyper om Grønland og Arktis. København: Forlaget Bios, p. 29.

Box, J.E., Yang, L., Bromwich, D.H. and Bai, L.S. (2009). Greenland Ice Sheet surface air temperature variability: 1840-2007. *Journal of Climate*, 22, 4029–4049.

Brincker, B. (2017). Images of the Arctic: Visualising Greenland as an indigenous people and a modern nation, *Visual Studies*, 32(3), 251–261.

City Population (2019). Population in Greenland, Available http://citypopulation.de/Greenland.html (accessed 14 July 2019).

Cogley, J.G., Hock, R., Rasmussen, L.A., Arendt, A.A., Bauder, A., Braithwaite, R.J. et al. (2011). Glossary of Glacier Mass Balance and Related Terms. *IHP-VII Technical Documents in Hydrology* No. 86, IACS Contribution No. 2, UNESCO-IHP, Paris.

CREDE (Center for Research on Education, Diversity & Excellence) (2019). The CREDE Five Standards for Effective Pedagogy. Available https://manoa.hawaii.edu/coe/credenational/ (accessed Jan 20th, 2019).

Cuffey, K.M. and Paterson, W.S.B. (2010). The Physics of Glaciers. Elsevier Science, Amsterdam, p. 704.

Danish Agency for Higher Education (2014). Guidance in Education – The Educational Guidance System in Denmark. 4th edn. Denmark, Euroguidance Denmark.

de Wit, C.A., Balmer, J., Muir, D.C.G., Vorkamp, K. and Wilson, S. (2019). Chemicals of Emerging Arctic Concern: Preface. Emerging Contaminants Bind 5 2019 s. 1–3.

Demmert, W.G. and Towner, J.C. (2003). A review of the research literature on the influence of culturally based education on the academic performance of Native American students. Northwest Regional Educational Laboratory, Portland, OR.

Direktoratet for Kultur, Uddannelse, Forskning og Kirke (2004). Kulturredegørelse for Landstinget: Rundt om kulturen, Grønland som kulturnation. Retrieved February 21, 2019, from the Home Rule's website: http://dk.nanoq.gl/Emner/Landsstyre/Departementer/Departement_for_kultur/Kultur/~/media/0E06EC4605424DDA80E05E22E006A21F.ashx.

Elkington, J. (1994). Towards the Sustainable Corporation: Win-Win-Win Business Strategies for Sustainable Development, *California Management Review*, 36(2), 90–100.

Elmore, R.F. (2006). School Reform from the Inside out. Policy, Practice, and Performance. Harvard Education Press, Cambridge Mass.

ESA CCI (2019). Download satellite data on surface elevation change, ice velocity, calving front location and grounding line location. European Space Agency – Climate Change Initiative. Available http://www.esa-icesheets-cci.org and http://cryoportal.enveo.at/data/ (accessed December 20th, 2018).

EU Commission (2013). Study to evaluate the performance of Higher Education in Greenland. EU Commission Report, Brussels.

EU Commission (2014). Programming Document for the Sustainable Development of Greenland 2014-2020 (2014/137/EU) Brussels.

Fausto, R.S., van As, D., Box, J.E., Colgan, W., Langen, P.L. and Mottram, R.H. (2016). The implication of nonradiative energy fluxes dominating Greenland ice sheet exceptional ablation area surface melt in 2012. *Geophys. Res. Lett.* 43, 2649–2658. https://doi.org/10.1002/2016GL067720.

Government of Greenland (2009). Greenland Parliament Act of 7 December 2009 on mineral resources and mineral resource activities (the Mineral Resources Act) with amendments from Greenland Parliament Act No. 26 of December 18 2012. www.bmp.gl/ images/stories/about_bmp/0912-Lov-ommineralske-raastofferOVERSAT_2013_2.pdf (accessed 10 January 2019).

Greenland Home Rule (2000). En vision om fremtiden – Oplaeg til strukturpolitisk handlingsplan (A Vision of the Future – Outline for a structural policy plan). Greenland Home Rule, September 2000.

Greenland Home Rule (2003). Parliamentary Act n. 7, April 11, 2003: The gender equality Act.

Gulløv, H.C. (1979). Home Rule in Greenland. *Etudes/ Inuit /Studies* 3(1), 131–142.

Hovelsrud, G.K., Poppel, B., Oort, B. and Reist, J.D. (2011). Arctic Societies, Cultures, and Peoples in a Changing Cryosphere. *Ambio*, 40(1), 2.

Jennes, D. (1968). Eskimo administration. V. Analysis and reflections. *Artic Institute of North America (Montreal), Technical Papers*, 21, 1–47.

Joughin, I., Abdalati, W. and Fahnestock, M. (2004). Large fluctuations in speed on Greenland's Jakobshavn Isbrae glacier. *Nature*, 432(7017), 608–610. doi:https://doi.org/10.1038/nature03130.

Kahlig, W. and Banerjee, N. (2007). Børn o gunge i Grønland – en antologi. MIPI. Ilisimatusarfik. Milik Publishing.

Kleemann, N. (2018). Greenland in Figures 2018. Statistics Greenland.15th revised edn. Greenland: Statistics Greenland.

Kleist, K., Jacobsen, M., Nielsen, P., Jacobsen, M.T. and Knudsen, R.J. (2015). Everybody on board. The human dimension. A Greenland perspective on capacity building in the Arctic. Greenland. A Greenland Perspective Publication.

Langen, P.L., Fausto, R.S., Vandecrux, B., Mottram, R.H. and Box, J.E. (2017). Liquid Water Flow and Retention on the Greenland Ice Sheet in the Regional Climate Model HIRHAM5: Local and Large-Scale Impacts. *Earth Sci.* 4, 110. doi: https://doi.org/10.3389/feart.2016.00110

Larsen, J.N. and Fondahl, G. (2015). Arctic Human Development Report: Regional Processes and Global Linkages. Nordisk Ministerråd, Copenhagen, p. 500.

Lee, D. and Newby, H. (1983). The Problem of Sociology: An Introduction to the Discipline. Hutchinson, London.

Mankoff, K.D., Colgan, W., Solgaard, A., Karlsson, N.B., Ahlstrøm, A.P., van As, D., Box, J.E., Khan, S.A., Kjeldsen, K.K., Mouginot, J. and Fausto, R.S. (2018). Greenland Ice Sheet discharge from 2000 to 2018, *Earth Syst. Sci. Data Discuss*. https://doi.org/10.5194/essd-2019-29, in review, 2019. https://doi.org/10.5194/essd-2019-29

Marquardt, O. (2010). Income distribution in west Greenland in the second half of the nineteenth century. *In:* Karen Langgård, Flemming Nielsen, Birgit Kleist Pedersen, Kennet Pedersen and Jette Rygaar (eds) Cultural and Social Research in Greenland: Selected Essays 1992-2010. Nuuk, Greenland: Ilisimatusarfik, pp. 55–78.

Marsh, C.J. (2009). Key Concepts for Understanding Curriculum. 4th edn. Routledge, Oxford.

NORSOK (2010). Z-013 Risk and Emergence Preparedness Assessment. NORSOK, Lysaker Norway.

OECD (2011). Guidelines on multinational enterprises (2011 Edition). Organisation for Economic Co-operation and Development Publishing, Paris. Available www.oecd.org/corporate/ mne/ (accessed November 12th, 2018).

Oltmanns, M., Straneo, F. and Tedesco, M. (2018). Increased Greenland melt triggered by large-scale, year-round precipitation events. *The Cryosphere Discuss*. https://doi.org/10.5194/tc-2018-243.

Poppel, B. (2006). Interdependency of subsistence and market economies in the Arctic. *In:* Solveig Glomsråd and Iulie Aslaksen (eds.). *The Economy of the North*. Statistisk Sentralbyrå/Statistics Norway, pp. 65–80. www.ssb.no/a/english/publikasjoner/pdf/sa84_en/ kap5.pdf.

Poppel, B. (2015). SLiCA: Arctic living conditions – Living conditions and quality of life among Inuit, Sami and indigenous peoples of Chukotka and the Kola Peninsula. Nordic Council of Ministers, Denmark, p. 67.

Poppel, M. and Chemnitz, J.K. (2009). Køn og magt I politik og erhvervslivet I Grønland. *In:* Niskanen, K. och Nyberg, A. (ed.), *Kön och makt I Norden. Del l Landsrapporter*. TemaNord 2009:569. Nordiska ministerrådet, Köpenhamm 2009, pp. 341–358.

PROMICE – Programme for Monitoring of the Greenland Ice Sheet. (2019). Danish initiative to monitor and assess changes in the mass budget of the Greenland ice sheet. http://www.promice.org

Rivera, H. and Tharp, R.G. (2006). A Native American community's involvement and empowerment to guide their children's development in the school setting. *Journal of Community Psychology*, 34(4), 435–451.

Ryan, A. (2011). ESD and Holistic Curriculum Change, p. 3, Available http://www.heacademy.ac.uk/resources/detail/sustainability/esd_ryan_holistic (accessed 12 December 2018).

Sørensen, A.K. (2006). Denmark-Greenland in the twentieth century. Meddelelser Om Grønland. 34. The Commission for Scientific Research in Greenland, Copenhagen.

Spindler, G.D. (1997). Education and Cultural Process. Anthropological Approaches. 3rd edn. Stanford University and Waveland Press, Long Grove, Illinois.

Statistics Greenland (2019). Available http://www.stat.gl/dialog/topmain.asp?lang=en&subject=Education&sc=UD#chart-wrapper (accessed 20th January, 2019).

SWIPA (2008). Climate Change and the Cryosphere: Snow, Water, Ice and Permafrost in the Arctic (SWIPA). Project description as approved by Senior Arctic Officials in Svolvær.

Thomsen, H. (1996). Between Traditionalism and Modernity. *In:* Cultural and Social Research in Greenland 95/96: Essays in Honour of Robert Petersen. Ilisimatusarfik and Atuakkiorfik, Nuuk, pp. 265–278.

Thomsen, H. (1998). Ægte grønlændere og nye grønlændere – om forskellige opfattelser af grønlandskhed (Real Greenlanders and new Greenlanders – about different perceptions of Greenlandicness). *Den Jyske Historiker*, 81, 21–69.

TMECCGE (2014a). The Education Plan II of the Government of Greenland, April 2014. The Ministry of Education, Church, Culture and Gender Equality, Government of Greenland, Nuuk.

TMECCGE (2014b). Education the Key to Our Future, October 2015. The Ministry of Education, Church, Culture and Gender Equality, Government of Greenland, Nuuk.

TMECCGE (2015) Education the Key to Our Future, October 2015. The Ministry of Education, Church, Culture and Gender Equality, Government of Greenland, Nuuk

UNESCO (2005). United Nations Decade of Education for Sustainable Development (DESD) 2005-2014. UNESCO, Paris, p. 2.

United Nations General Assembly (1962). Resolution 1803 (XVII) of 14 December 1962, Permanent sovereignty over natural resources. United Nations, New York.

United Nations General Assembly (2015). Transforming our world: the 2030 Agenda for Sustainable Development. Resolution adopted by the General Assembly on 25 September 2015, A/RES/70/1, 4th plenary meeting 25 September 2015, United Nations. New York.

Vygotsky, L.S. (1978). Mind in society: The development of higher psychological processes. Harvard University Press, Cambridge, MA.

WCED (World Commission on Environment and Development) (1987). Our Common Future. Oxford University Press, Oxford, pp. 12–14.

Wyatt, T.R. (2009). The role of culture in culturally compatible education. *Journal of American Indian Education* 48(3), 47–63. https://jaie.asu.edu/sites/default/files/483_2009_3_wyatt.pdf.

Wyatt, T.R. (2010). Supporting an externally developed Model of education in Greenland. *International Journal of Educational Reform*, 19(3), 156–171. https://doi.org/10.1177/105678791001900301.

Wyatt, T.R. (2012). Atuarfitsialak: Greenland's cultural compatible reform. *International Journal of Qualitative Studies in Education*, 25(6), 819–836. https://doi.org/10.1080/09518398.2011.558033.

Chapter 11
Education for Sustainable Development (ESD) in Tourism-Oriented Vocational Degree Programmes in Turkey

Emine Cihangir and Mehmet Şeremet

1 Introduction

Tourism is a trillion dollar global industry and accounts for 10.4% of the total GDP (WTTC 2018). The sector supports 313 million jobs, perhaps 1 in 10 jobs worldwide, and creates a tourism community involving local people, the sector, NGOs, policy-makers and, of course, the education sector (Knusden et al. 2007; Haigh 2006). However, tourism, as a term, has gone through many changes in the last three decades, many driven by the paradigms of neo-liberalism and post-modernism (Xu et al. 2014). In the early development stage, the penetration of tourism into society was mainly controlled by 'authenticity', which was later turned into a consumption-oriented economy guided by neoliberal concepts such as commercialisation and commodification (West and Carrier 2004). As it evolved into a highly profitable sector in the economy, emerging economies seized upon the development of tourism as a means of revitalizing local economies, redressing regional imbalances, and realising sustainable economic benefits.

As a rapidly developing county, from the mid-1950s, Turkey became involved in tourism through its riviera and its quintessential tourist attractions. This resulted in an influx of tourists, mainly from "Western" countries that were already involved in the tourism industry and whose domestic tourism industries now sought new destinations. Hence, these mainly Western visitors engaged with tourism activities that had been planned by their home tour operators and local chain hotels (Zimmermann 2018). Later, the search for authenticity in the country was tempered by the mass-tourism-oriented industrial encroachment into coastal areas. In this massification

E. Cihangir (✉)
Department of Tourism Management, School of Tourism and Hotel Management, Van Yuzuncu Yil University, Van, Turkey

M. Şeremet
Department of Transportation and Business Engineering Management, Maritime Faculty, Van Yuzuncu Yil University, Van, Turkey

E. Pénzesné Kónya et al. (eds.), *Environmental Sustainability Education for a Changing World*, https://doi.org/10.1007/978-3-030-66384-1_11

period, formerly untouched seaside towns became commercialised and, from an authenticity point of view, 'spoilt' except for a few formally protected areas and places where the local people, unwittingly, became tourist attractions themselves. When the country, finally, confronted the concept of sustainable development after the Rio summit in the early 1990s, it was far too late to protect the Turkish Riviera from counter-productive urban sprawl, witness Antalya and surrounding locations, which have a huge involvement with the international sea-sun-sand tourism market.

However, Turkey has many unique tourism assets and, thus, deserves both an effective protection plan and sustainability policy. It is probably true to say that Turkey's environmentalist movement began with the process of applying for full membership of the European Union (EU). When the European Union's *European Charter for Sustainable Tourism in Protected Areas* was adopted, this was the first concerted attempt to set out regulations and principles for sustainable development (Wallace and Russell 2004). One of the principles aimed at developing rudimentary green skills amongst tourist visitors (Sadorge 1998). However, this did not develop a climate of sustainability in the country.

Instead, the idea of sustainable development (SD) underpinned the emergence of Turkey's sustainable tourism industry, which, initially, remained small scale and under-informed. Perhaps, no one anticipated that this might result in counter-productive impacts on so far unspoiled landscapes and their indigenous cultures. One reason was that the professional cadre of tourism developers was mainly transferring experience from mass tourism and its facilities, such as chain hotels and restaurants. Eventually, an alternative tourism "spin-off" emerged from the sustainability agenda producing "niche" tourism products. However, as many tourism researchers underline in the literature, some concepts (e.g. authenticity, tourist gaze) were coined by the western-oriented perception and, probably, this is the case for SD (sustainable development) too (Katayama et al. 2018). This might prove that similar concepts can lead to misunderstanding by non-western societies experiencing different phases of socio-economic development. Hence, Turkey tends to interpret the SD concept in a negative sense, thinking that an SD policy might challenge the country's future economic development rather than defend its future viability. Of course, SD encompasses a variety of themes including community wellbeing, economic development and the protection of the environment. Indeed, the tourism sector offers a reflection with the introduction of Ecotourism initiatives that dovetail with the major aims of the United Nations Sustainable Development Goals (SDGs) in tandem with its subsequent community-based and environmental-based structures (Weaver and Lawton 2007). However, poor understanding of the concept amongst the Turkish tourism community led to the belief that sustainability involved creating a nature-based tourism market rather than investing in sustainable energy and lessening the poverty and inequalities of rural communities or encouraging the empowerment of the locals through providing employment and markets for their produce. Instead, this limited vision risks hindering the integration of the country in the sustainability era and overlooks the real benefits for long term national development to be had from investing, systematically, in sustainable tourism rather than permitting tourism assets and environments to become devalued and degraded by poorly conceived development.

Özden (2012), reflecting on failures in earthquake disaster management, discusses the problems of transitioning Turkish society from passive fatalism to something more proactive and rehabilitative in orientation. There are two salient matters in this discussion. One is raising awareness; the second is planning ahead, which in this case means promoting relevant skills in the developers and practitioners of the Tourism Industry. Inevitably, this is a task for education, which is probably one of the reasons that UNESCO launched the international decade of ESD (Education for Sustainable Development) (Higgitt 2006). While NGO-supported informal education might canvass the UN's later SDGs to society in the name of 'sustainability literacy', real progress depends on developing these ideas through formal vocational education and training. This needs a strong policy that might also help the tourism sector to shift its emphasis from the profit-oriented present to the SD-oriented future (Boley 2011). The need for this transformation is a sufficient reason for introducing greening policies into the tourism business sector; there remains a need for the training of existing, as well as future, office workers, hotel managers and relevant staff who are working in or entering the tourism sector (Boley and Perdue 2012). There is, therefore, also a vitally important need for a reciprocal relationship between the tourism industry and higher education. This chapter, therefore, aims to provide critical insight into Turkey's approach to introducing sustainability thinking into Turkey's tourist industry as well as into the extent to which higher education has made progress in teaching ESD within tourism-oriented degree programmes.

To date, the bulk of ESD-oriented research in Turkey has focused on issues dealing principally with teacher training for the school sector, although, as Haigh (2006) emphasises, this may well be a useful route towards helping ESD penetrate into Higher Education Institutions (HEIs). Meanwhile, the vast bulk of ESD-oriented academic work in the field involves "attitude-scale-development" type studies in the education faculty and secondary schools (Alkis 2008; Katayama et al. 2018).

By contrast, the mainstream of research in the tourism field has focussed on the relationship between the environment and tourism; part of an internationally burgeoning trend that began with the foundation of the *Journal of Ecotourism*, *International Journal of Sustainable Tourism Development* and *Tourism Geographies* (Gough and Scott 1999). By comparison, in Turkey, reflection on the relationship between tourism and environment has been somewhat limited. For example, a review of the tourism-related theses on the principles of SD found that this research theme is weak and only began to make a small presence in the mid-2000s.

Nevertheless, Turkey's situation reflects the problems of tourism research in general with its unhelpful divide between theoretical and applied work and shortage of integrated applied problem-solving research, not least within the sustainability-oriented tradition (Butler 2015). Today, a small number of SD-related tourism studies discuss how tourism has a potent effect on the values and beliefs of the tourist (see Brown 2013; Marvell and Simm 2016; Simm and Marvell 2015). However, this does suggest how tourism-based sustainability education might be orchestrated through a more general discussion of sustainable development that aims to help reconcile the conflicts between the demands of preserving the environment, cultural resources and tourism development.

2 Problems Facing the Development of Education for Sustainable Development (ESD) in Turkish Higher Education

Since the early 1990s, while the aim of the prosperous countries has been to create enterprise economies, a question remained to be asked, which is how this development will be achieved in the context of sustainability (Higgitt 2006). Thanks to the efforts of the United Nations (UN) and UNESCO, in the early 2000s, a list of suggestions emerged. On the top of this list is 'Education for Sustainable Development' (ESD) (UNESCO 2003). Perhaps not surprisingly, countries such as the USA, Australia, New Zealand, and the UK have made great progress in the integration of this subject into their systems (Chapman et al. 2006; Bednarz 2006; Scott and Gough 2006). However, the ESD that has been built has been constructed according to a Westernised vision of Environmental Education. This vision may not be appropriate for developing countries, where economic development and the creation of employment have to remain core goals for the community. For example, in the ratification of the Climate Change Agreements (CCA) first framework in New York, Turkey was initially placed with the industrialised countries, which were required to reduce their greenhouse gas emissions by 20% before 2010. Soon, it was recognised that this could, potentially, nullify the economic development of Turkey. So, the Turkish government proposed a change at the meeting Marrakech of Morocco in 2001 to move into the list of developing countries, where no action was required (Türkeş 2001). Probably, this would have also been a reason for the country not to make any progress regarding environmental issues, despite the fact that Turkey's delegation was involved in political action for ESD (Katayama et al. 2018). Most importantly, the Turkish Government's action plan committee did not include representatives from the Ministry of Education during the development of the action plan in 2003 (Köse 2018). Ten years later, in 2013, this policy was changed. Finally, the offices contributing to the mitigation of Climate Change included the Ministry of Education, just in time for the final year of the UN's Decade for ESD (2005–2014).

Meanwhile, in 2005, the school curricula of both primary and secondary schools were revised and, subsequently, environment and society became one of the learning areas in both the Social Studies and Geography curricula. This learning area addresses some major issues related to sustainable development. However, it does not engage with sustainable tourism; instead, the focus is education. Mainly, the degradation of the environment and pollution caused by human activities (Yaşar and Şeremet 2009).

One important feature of the Turkish HE system is its lack of autonomy, which means that the current HE structure prevents many universities from playing to their individual strengths. The Higher Education Council of Turkey (HECoT) sets the common policies of Turkey's HEIs centrally (Seremet and Chalkley 2012). This makes it difficult for individual HEIs to respond quickly to the needs of rapidly changing knowledge-based local (and global) economies. Consequently, few of the state universities have a unique perspective on sustainability.

In recent years, HECoT has proposed a new approach that may allow some universities to play to their strengths, although, of course, this 'strength' would be assigned centrally by the associated commission within HECoT. For example, one HEI (Sinop University) was chosen in 2016 as the university to lead the development of sustainable environmental management technologies, ecosystems and sustainable built environment. However, this does not necessarily mean that this university will offer an ESD-based education or even that it is best suited for this role. This university, for all its many qualities, was not ranked within the green league of international universities. Indeed, research universities, such as ITU, METU, Ankara, and Erciyes have proved far more successful in providing sustainability education and developing a green campus (limiting energy and water resource wastage). This situation might support the argument that the Higher Education Council of the country has no effective programme for greening education, despite internationalism and employability becoming priorities for the Universities in Turkey in recent years (Şeremet 2015, 2017).

Nevertheless, a few HEIs, mainly non-profit universities, have become involved in global university greening initiatives (e.g. IAU; UI Green Metric; HESI) and presented their individual arguments for being environmental-friendly. For example, the Green Metric University Initiative's 2019 ranking of world universities lists 30 HEIs from Turkey including Istanbul Technical University, which was listed in the top 100. This, particularly, shows that sustainability in the curriculum and teaching of the engineering programmes played an important role in this ranking. For example, Bogazici University has a service-driven vocational Masters programme in Sustainable Tourism Management and includes the greening issue in institutional development policy (Katayama et al. 2018). Katayama et al. (2018) find that engineering programmes offer one-fourth of the environmental-driven modules in the country. It is, therefore, not surprising that, more recently, the research universities of Turkey achieved good scores within the global greening ranking.

Civic society is an important dimension in the Sustainability agenda, NGOs (e.g. TEMA, TURMEPA, TURÇEK and ÇEV) in Turkey have become involved in dozens of awareness-raising projects pertaining to ESD and many took private institutions as partners (cf. Haigh 2006; Alkis 2008). Although this is slightly encouraging for Turkish Higher Education, most HEIs prefer not to become involved in any independent environmental actions and so neglect this opportunity to raise awareness among those who are, potentially, the future professionals of the country (Wade 1999). This might, equally, be true for the tourism management and hospitality programmes in the Turkish HEIs.

3 The Structure of Tourism-Oriented Degree Programmes in Turkey and Their Relationship with Employability and ESD

Tourism is becoming a popular vocational subject in Turkey. There are a total of 133 tourism-oriented degree programmes (e.g. Tourism Management, Hotel Management, Tour Guidance and their variants), of which 88 are based in larger academic faculties and the remainder in specialist Tourism Schools. In HEIs, vocational-oriented tourism programmes appear in two contexts: the first is as an independent school such as a Tourism & Hospitality School and/or Tourism Faculty (recently, the more common), the second is as programme-based provision, which is placed in Business Schools/Faculties and which provides about 10% of tourism-oriented degree programmes in Turkey. The first type tends to offer a relatively 'holistic' mix of modules both from the business side (e.g. marketing, business, economy and accounting) and from the more programme-specific socio-cultural-oriented aspects of reception management, customer relations, hotel management, human-resources and management strategies, and leadership. The latter approach, which is becoming more widespread, aims to engage people more with the set of public-facing business skills they will need in practice. Typically, programmes offer major two-year upper-level modules that fit more generally into the Business and Marketing programmes and, later, discipline-specific modules that enable them to gain professional entry to the tourism field.

Given that sustainable tourism is a growing sector and receives lavish attention from 'eco-people', tourism-oriented undergraduate programmes might have been expected to respect the sustainability agenda (Boley 2011). However, as Sterling (2004) posited, ESD should be an integral, holistic pedagogic approach to the teaching of subject teaching rather than being just a title line on the syllabus. This illuminates the fact that SD is not simply collateral knowledge to be attached to the formal curricula. Rather, it is a state-of-art pedagogy including a multi-dimensional system that covers the wider perspectives of environmental education (Gough and Scott 1999, 2003) and that facilitates students' learning with engagement in greening issues through active projects on local and regional sustainability issues (Chalkley 2006b). Given the nature and breadth of tourism, the programme is highly suited to this kind of development (Boley 2011). However, tourism has a twofold tradition, which arises from the nature of tourism knowledge, namely partly business-oriented and partly socio-culture oriented (Aramberri 2010; Liburd 2012). Hence, tourism must support skills development in the business field as well as graduate employability skills specific to the tourism sector (Cihangir and Şeremet 2018; Brown 2013; O'Reilly 2006; Inkson and Myers 2003). Inevitably, there develops a tension between the contrasting world views of promoting immediate profitability and promoting long term sustainability. In Tourism's HEI curricula, these two contrasting views can also collide, pitting employability concerns against ESD, for example in the UK and Australia (Jones et al. 2008; Wilson and von der Heidt 2013).

Arguably, this challenge has been handled well by some Geography programmes, especially those in the UK that have 'championed' pedagogic and curriculum innovation. The incorporation of ESD in the UK's Geography programmes was also encouraged centrally by the changes in the formulation of its 'Subject Benchmark Statement', which was published by the UK Government's QAA (Quality Assurance Agency for Higher Education) in 2007 (Seremet 2013). The UK's 'Subject Benchmark' documents are drawn up as statements of the aims, purposes and scope of the discipline, what the discipline covers and what and how it should be taught. Despite Geography being an academic rather than applied discipline like Tourism, Turkey's tourism programmes might draw lessons from such experience, perhaps by creating their own idealised benchmark for the discipline, or less formally, by addressing what might be the qualities or attributes they might collectively wish to develop in all of their graduates.

3.1 ESD Provision Within the Tourism-Oriented Degree Programmes

Principally, Turkey's tourism programmes aim to deliver content that is directly relevant to the hotel and tourism sector, although 'greening' is widely accepted (Korkmaz and Atay 2016; Gürlek and Tuna 2018; Hüseyinli and Küçükkancabaş Esen 2018). Higher education recognises only mainstream expectations of the sector rather than those of any "niche" markets. Tourism and hospitality programmes have therefore been dominated by business-oriented modules such as management, business, economics, accounting, statistics, service sector experience, entrepreneurship, and so on. Only students with interest in studying, for example, alternative tourism enrol in such elective modules.

When looking at undergraduate level ESD provision in Turkish Tourism Schools/Faculties, it is clear that there are two types of provision: "wider" Sustainable Development modules (e.g. Tourism Geography, Tourism Sociology, Social Responsibility Project, Entrepreneurship and Gender Equality etc.) and "Focus" Modules (Sustainable Tourism, Sustainable Tourism and Development, Management of Sustainable Tourism, Tourism and Environment, Sustainable Environment and Tourism etc.). Almost all Tourism Schools/Faculties have at least one ESD-type course, which shows that the idea of SD is accepted by most undergraduate courses. The number of wider modules also seem to be high; meaning that each department has at least one (Table 11.1). Additionally, these wider modules are not mutually exclusive; as their curricula deal with themes related to sustainability principles such as equality, social responsibility, SD-based planning, cultural diversity, and tourism's social and economic effects on environment, place and culture. However, these modules in Turkey commonly last for a single term, meaning that these are shorter and carry less credit than typical modules elsewhere (Seremet 2013). The main exceptions are modules that include work-based training.

Table 11.1 The provision of ESD within tourism-oriented programmes by the type of school

	Keywords of the module names in the tourism-oriented programmes							
	"Focus" Modules				"Wider" Modules			
	SD	Environment	Green Marketing/Management	Alternative Tourism	Gender Equality	Social Responsibility	Tourism Geography/Tourism Sociology/Tourism Psychology	Entrepreneurship
Faculty	53	39	2	14	8	15	160	81
Schools	25	21	2	15	7	13	63	44
Total	78	60	4	29	15	28	223	125

Of course, ESD is not only a matter of syllabus. It needs to be supported by an appropriate, effective, pedagogy of delivery (Sterling 2004). This relates, particularly, to skills development, which is also important for the employability agenda. In Turkey, ESD coverage in the Tourism curricula seems to be very patchy, not progressive, and probably not involving any of the pedagogical approaches considered relevant to ESD-oriented modules, such as project-based learning; problem-solving and experiential learning. The exceptions are within engineering programmes, where such pedagogies are applied to sustainability issues and where students are, routinely, exposed to hands-on learning activities and project-based teaching, which supports the holistic vision of sustainability themes such as environment, development and social equality. Katayama et al. (2018) point out that the green provision of Turkish HEIs teaches students the skills of coping with environmental problems and that this provision is based on the 'liquid' theory of environmental education. As many studies have emphasised, sustainability education should be based on practical approaches. However, according to the authors' experiences, the programmes in the social disciplines embracing vocational-oriented degrees tend to be delivered lecture style. Tourism degree programmes may require less technical and academic infrastructure than most engineering-driven programmes. However, its pedagogy lags behind those of the leading engineering-oriented degree programmes.

3.2 *Employability Skills and ESD in Tourism-Oriented Programmes*

Turkey's tourism programmes might draw lessons from the experience of Geography in the UK, where ESD pedagogy is more advanced than in most other undergraduate programmes. The country is probably a pioneer in ESD (Chalkley 2006a; Cotton et al. 2007; Jones et al. 2008), not least because of former Government subsidies for the development of subject-based 'Centres of Excellence in Teaching and Learning', some of which gave a priority to ESD at HE level. These ESD activities blended well with an employability agenda, which had been important since the UK Government's Dearing Report of 1997 (Dyer et al. 2006). These goals were widely adopted by Geography departments (Chalkley 2006a; Selby 2006; Cotton et al. 2007; Jones et al. 2008; Schweinsberg et al. 2013) and to lesser extent those in cognate disciplines (Fenner et al. 2005; Jones et al. 2008). ESD's contribution to the development of both the transferable and core skills, (i.e. employability skills) of university graduates dovetailed neatly with these Geography programmes teaching of ESD (Gress and Tschapka 2017). Probably, tourism's vocational-degree programmes share aspects of Geography's nature and breadth. However, tourism programmes also develop practical skills. Among the several inherent virtues of business and management education is the building of skills that are intrinsically valuable in the job market. For these reasons, there is no reason why Tourism cannot both learn from, adopt and even advance beyond Geography's successful incorporation of ESD and SDG sustainability goals within its curricula. After all, ESD is defined as "a process of learning how to make decisions that consider the long-term future of the economy, ecology and equity of all communities and as being about capacity building for 'future-oriented thinking'" (UNESCO 2003, p. 3). Given the employability and skills development role of the Higher Education sector, it makes sense to embed the forward-thinking of ESD in a larger employability agenda (Gress and Tschapka 2017; Jones et al. 2008). Turkey's National Higher Education Qualification Framework (NHEQF) recognises that there are key personal attributes which the students should attain before graduation. These include the understanding of social rights and social justice, awareness of environmental problems and the protection of cultural values (HECoT 2012). Graduates who work in the service sectors such as tourism, hospitality, recreation and guidance need these skills and also skills such as intercultural communication and a sense of social responsibility for people and the environment. Although, officially, these attributes seem to be only advisory, the question remains of how these essential ESD-oriented skills can be embedded into the curriculum of vocational-oriented programmes, particularly when they are not commonly developed by other degree programmes in the country.

To date, little attention has been paid to the development of employability skills in the context of ESD, the literature focuses principally on the sustainability-based managerial skills, (inter-cultural) communication skills, problem-solving (Boley 2011; Sheldon et al. 2011; Hatipoglu et al. 2014) and, sporadically, ICT skills (Ali and Frew 2013; Şeremet and Chalkley 2015). Among these, a few ICT-oriented skills are gaining traction in the field of sustainable environmental management, which is an important facet of the current tourism sector. Many of these technical quantitative skills matters are outside the scope of Tourism-oriented vocational degree programmes in Turkey. However, Geographical Information Systems (GIS) is, potentially, important for both tourism and sustainability education (Bahaire and Elliott-White 1999; Ho and Ang 2018; Hwang 2013; McAdam 1999; Ali and Frew 2013; Şeremet and Chalkley 2016), not least because it offers a set of key tools for management, planning and its representation that are particularly important for sustainable tourism management and tourism planning. However, while Ali and Frew (2014) assert that tourism sector managers use the GIS to sustainably manage the tourism destinations, none of the tourism programmes in Turkey include this subject. In Turkey, the GIS job market has been mainly occupied by the graduates of Geography and Engineering (e.g. Environment, Geology and Architecture and Planning) (Şeremet and Chalkley 2016).

Turkey, and many other countries in the Mediterranean basin, has suffered from the over-development of sea-sun-sand tourism. Nevertheless, the current mass tourism infrastructure has great scope to adopt the principles of SD-oriented management (e.g. empowerment of the locals by employment, local products, energy use etc.) and so cause less damage to their environment. Unfortunately, training and education appropriate for this task is not very common in Turkish tourism programmes. The few existing modules, such as green marketing and management modules, are mainly offered by the non-profit universities, which tend to be more business and tourism sector-oriented in their undergraduate provision. Hence, training in sustainability-based management, communication and problem-solving skills is something that the State HEI sector, as a whole, needs to develop. As in the case of GIS skills, technical issues such as energy-saving, recycling of waste and green energy-use are considered too technical for most tourism programmes to include in addition to training the managers for this field. Hence, the technical managers of hotels are mainly chosen from graduates in electronic, electrical and environmental engineering. Nevertheless, currently, none of the sector's representations suggests that HEIs should create skills-driven modules that might address these skills shortages in the sector.

Empowerment of social groups is, also, of relevance to the sustainability agenda. In this context, gender equality is one of the Sustainable Development Goals (SDGs) that emerged from the United Nations' Conference on Sustainable Development Summit in Riode Janeiro (2012) (Sachs 2012). This issue is also, gradually, becoming an agenda issue in Turkey, where patriarchalism is a ubiquitous feature of society. A joint project between UNESCO and Turkey's Ministry of Education was implemented between 2007 and 2012 towards alleviating gender inequality in Turkish society. Later, HECoT became involved in the project, signing a memorandum, which was announced on their website. However, HECoT, reacting to issues raised in society, has, recently, stepped back from this decision (BBC 2019). Reflection of this can be seen easily in the gender equality statistics of the World Economic Forum (WEF), where Turkey was ranked at number 121 of 136 countries (WEF 2018). Today, female workers represent only 20% of the workforce in the Tourism and Recreation industry. Hopefully, ESD and the development of sustainable tourism may help Turkey to do more to support the SDGs that advocate increasing women's participation in employment (Cukier 2002; Gorg 2000; Hashemkhani Zolfani et al. 2015).

Compulsory work-placement in tourism-oriented degree programmes might be one of the ways of addressing this challenge. Education might, thereby, enable widening the participation of tourism students through work-placements that placed students, temporarily, in tourism-oriented jobs in other parts of the country. Considering the fact that tourism development is evenly distributed across the larger territory of the country from eastern border cities (e.g. Van, Şırnak, and Ağrı) to the western resort towns (e.g. Antalya, Mugla and İzmir provinces), compulsory work placement modules might create a promising opportunity for female tourism students to experience different cultures and work environments. This could, potentially, be an important part of their personal development and it also supports widening participation in higher education and enhancing women's participation in the sector.

Entrepreneurship is another important skill for tourism graduates who are familiar with economic and management subjects as well as environmental issues. Encouraged by a Ministry of Industry initiative, today, many departments offer entrepreneurship modules that develop the skills of students from disadvantaged backgrounds, who might later opt to establish their own business or to undertake social projects that help support local communities and local products. In the accommodation sector, for example, a boutique hotel sector is growing alongside an emerging "niche" tourism market supported by alternative tourism types (e.g. eco-tourism, rural tourism, winter tourism, adrenalin tourism, etc.).

Finally, regarding the distribution of the tourism degree provision across the country, this illustrates how, in recent years, the government has established many new HEIs in the east of the country, which traditionally has been poorer and had fewer HE opportunities (Şeremet and Chalkley 2016). So, although eastern towns like Van, Şırnak, Batman, Mardin, and Ağrı are quite distant from the country's main touristic attractions, many of these new eastern universities offer new tourism-oriented vocational degree programmes. The idea behind this is, perhaps, related to the desire to boost the tourism potential and infrastructure of eastern towns which, currently, are less developed and have fewer employment opportunities. Of course, there is still some way to go to realise this goal.

4 Conclusions and Recommendations for the Tourism Community

Tourism education is highly relevant to the goal of achieving sustainability in many different ways such as the empowerment of local communities, protection of the environment, and supporting the local economy. However, these positive changes can only be achieved with the active support of the tourism community and industry. In Turkey, the progress in the presence of Sustainable Development in tourism education is still rather limited and much more could be done. So, whilst it is accepted that a well-skilled workforce is an important underpinning of the rapidly growing Turkish tourism sector, the contribution to be made by university education in the next decades is still unclear. However, there is huge potential for enhancing employability and ESD. Today, advancing the quality of teaching and learning has been accelerated by creating individual aims for degree programmes within the NQF–HETR (the national degree qualification framework), although the adoption of these initiatives into programmes is very far from universal. Additionally, the conservative structure of the HE sector and staffing issues prevent HEIs from making great progress on these issues. So, despite, sometimes because of, its substantial bureaucratic system, Turkey has experienced a poorly planned and unsystematic growth of tourism education in the HE sector. Currently, there are 133 degree programmes and this may be one reason that the students' enrolment to some of the new programmes is declining.

However, internationally, tourism is a major economic growth area. International tourism has long been linked with unsustainable and destructive practices but, today, it is trying to improve its record and deliver sustainable development for its existing markets as well as developing green alternative tourism and green hotel management. There is, of course, the important issue of how this can be delivered in countries that have no substantive sustainability policy and strategy. HEIs, therefore, need to take responsibility to close the gap between the current attitudes and expectations within the sector and those of international society. This short review of the educational goals and structures of vocationally-oriented tourism programmes in Turkey reveals that while there is a strong provision of environmental education within the relevant curricula, much more needs to be done to boost ESD teaching through the application of the relevant pedagogic approaches that underpin sustainability-oriented skills development.

Probably, today, in Turkey's HE sector, the non-profit universities are 'one-step-ahead' on sustainability issues compared to the State universities, which have much stronger engineering and technical programmes. These, of course, are also important to the development of Education for Sustainable Development. However, there is still a gap between the needs, present and future, of Turkey's tourism sector and what is being delivered within the HEI programmes. Therefore, this section concludes with a set of suggestions, 'ESD-homework', which might offer some guidance to the major stakeholders in the sector.

For the practitioners in the tourism sector:

- Greening and sustainable management issues need to be discussed within the local HEI programmes, which are now very common across the country.
- A Masters programme in Sustainable Management of Environment/Hotels/Destinations is now offered by two departments and enrolment by both managers and workers should be encouraged.
- Engagement with the local communities to create a local supply chain from local products and so support their economy, should be harnessed to help "safeguard" the environment.
- Gender equality is important for ESD, so each hotel should train staff to be aware of the issues involved. Of course, some chain hotels are already offering this type of programme to their workers (e.g. "he for she").
- Providing specific employability quotas for local workers from vulnerable backgrounds are recommended.
- Sustainability skills should be reinforced and extended through the provision of Continuing Professional Development (CPD) programmes and projects with the tourism community.

Given that practitioner's direct experience is relevant to Tourism graduates, lecturers and programme coordinators in Tourism Schools should engage with such sources and draw lessons from the case studies;

- The importance of ESD should be recognized by all Tourism departments. This is important partly because ESD, as a means of raising students' awareness and developing students' general sustainability credentials, will be important for the future of the expanding tourism sector and also for preserving each location's touristic assets.
- For these reasons, ESD should be an essential component of every progressive syllabus. Sustainability awareness and some rudimentary ESD-oriented skills should be a fundamental graduate attribute for every Tourism student. There

should also be more opportunities at the Masters level to study Sustainability in Tourism, which might also open a path to deeper collaboration between the commercial sector and the HEIs.

- Key technologies, such as Geographic Information Systems (GIS), are gaining importance and might be introduced as open elective courses for the students in the later stages of undergraduate education.
- Each department might also consider creating open specialist modules that allow students to do projects and practice skills that are relevant to the SDGs and the sustainability of the Tourism Sector.
- Departmental/faculty teams should liaise with sector representatives (including alumni) when reshaping the curriculum of tourism programmes, a process which might bring valuable insights on current sustainability issues.

More generally, Tourism programmes might reconsider their roles and change away from being isolated business-driven curricula towards being multi-disciplined skills-driven programmes. All in all, it is, of course, fully acknowledged that some of these suggestions may face constraints such as budget and staff time. Moreover, in many cases, the proposed ideas may be difficult to achieve without support at the institutional or governmental level. Nevertheless, it is hoped that these recommendations will inform discussion and provide a more holistic agenda for change towards enhancing tourism students' learning and experience of ESD, so promoting graduates employability across the tourism sector.

References

Ali, A. and Frew, A J. (2013). Information and Communication Technologies for Sustainable Tourism. Routledge. London.

Ali, A. and Frew, A.J. (2014). Technology innovation and applications in sustainable destination development. *Information Technology & Tourism*, 14(4), 265–290. https://doi.org/10.1007/s40558-014-0015-7.

Alkis, S. (2008). Education for sustainable development in Turkey. *Internationale Schulbuchforschung*, 597–608. https://www.jstor.org/stable/43057364.

Aramberri, J. (2010). Modern Mass Tourism. London: Emerald.

Bahaire, T. and Elliott-White, M. (1999). The application of geographical information systems (GIS) in sustainable tourism planning: A review. *Journal of Sustainable Tourism*, 7(2), 159–174. https://doi.org/10.1080/09669589908667333.

BBC (2019). Eğitimde toplumsal cinsiyet eşitliğine karşı çıkan dernekler ne diyor? https://www.bbc.com/turkce/haberler-turkiye-47610198.

Bednarz, R.S. (2006). Environmental research and education in US geography. *Journal of Geography in Higher Education* 30(2), 237–250. https://doi.org/10.1080/03098260600717315.

Boley, B.B. (2011). Sustainability in hospitality and tourism education: Towards an integrated curriculum. *Journal of Hospitality & Tourism Education*, 23(4), 22–31. https://doi.org/10.1080/10963758.2011.10697017.

Boley, B.B. and Perdue, R.R. (2012). Destination management, competitiveness, and quality-of-life: A review of literature and research agenda. *In:* M. Uysal, R. Perdue and M.J. Sirgy (Eds.), Handbook of Tourism and Quality-of-Life Research, pp. 515–528. Springer, Dordrecht.

Brown, L. (2013). Tourism: A catalyst for existential authenticity. *Annals of Tourism Research*, 40, 176–190. https://doi.org/10.1016/j.annals.2012.08.004.

Butler, R. (2015). The evolution of tourism and tourism research. *Tourism Recreation Research*, 40(1), 16–27. https://doi.org/10.1080/02508281.2015.1007632.

Chalkley, B. (2006a). Education for sustainable development: continuation. *Journal of Geography in Higher Education*, 30(2), 235–236. https://doi.org/10.1080/03098260600717307.

Chalkley, B. (2006b). Tertiary teaching excellence: The CETLs scheme in England. *New Zealand Geographer*, 62(2), 155–157. https://doi.org/10.1111/j.1745-7939.2006.00067.x.

Chapman, D., Flaws, M. and Le Heron, R. (2006). A due diligence report on New Zealand's educational contribution to the UN decade of education for sustainable development. *Journal of Geography in Higher Education*, 30(2), 281–292.

Cihangir, E. and Şeremet M. (2018). Türkiye'de Turizmin Yapısal Sorunları). *In:* Ö.S. Doğan and E. Duran (Eds.), Türkiye Turizm Coğrafyası, pp. 343–380. Çanakkale: Paradigma Akademi.

Cotton, D.R., Warren, M.F., Maiboroda, O. and Bailey, I. (2007). Sustainable development, higher education and pedagogy: a study of lecturers' beliefs and attitudes. *Environmental Education Research*, 13(5), 579–597. https://doi.org/10.1080/13504620701659061.

Cukier, J. (2002). Tourism employment issues in developing countries: Examples from Indonesia. *In:* R. Sharpley and D. J. Telfer (Eds.), Tourism and Development, Concepts and Issues, pp. 165–201. Clevedon: Channel View Publications.

Dyer, A., Selby, D. and Chalkley, B. (2006). A centre for excellence in education for sustainable development. *Journal of Geography in Higher Education*, 30(2), 307–312. https://doi.org/10.1080/03098260600717406.

Fenner, R.A., Ainger, C.M., Cruickshank, H.J. and Guthrie, P.M. (2005). Embedding sustainable development at Cambridge university engineering department. *International Journal of Sustainability in Higher Education*, 6(3), 229–241. https://doi.org/10.1108/14676370510607205.

Gorg, H. (2000). Multinational companies and indirect employment: Measurement and evidence. *Applied Economics*, 32, 1809–1818. https://doi.org/10.1080/000368400425035.

Gough, S. and Scott, W. (1999). Education and training for sustainable tourism: Problems possibilities and cautious first steps. *Canadian Journal of Environmental Education (CJEE)*, 4(1), 193–212.

Gough, S. and Scott, W. (2003). Sustainable Development and Learning: Framing the Issues. Routledge.

Gress, D.R. and Tschapka, J.M. (2017). Bridging geography and education for sustainable development: A Korean example. *Journal of Geography*, 116(1), 34–43. https://doi.org/10.1080/00221341.2015.1119874.

Gürlek, M. and Tuna, M. (2018). Reinforcing competitive advantage through green organizational culture and green innovation. *The Service Industries Journal*, 38(7–8), 467–491. https://doi.org/10.1080/02642069.2017.1402889.

Haigh, M.J. (2006). Promoting environmental education for sustainable development: The value of links between higher education and non-governmental organizations (NGOs). *Journal of Geography in Higher Education*, 30(2), 327–349. https://doi.org/10.1080/03098260600717422.

Hashemkhani Zolfani, S., Sedaghat, M., Maknoon, R. and Zavadskas, E.K. (2015). Sustainable tourism: A comprehensive literature review on frameworks and applications. *Economic research-Ekonomska istraživanja*, 28(1), 1–30. https://doi.org/10.1080/1331677X.2014.995895.

Hatipoglu, B., Ertuna, B. and Sasidharan, V. (2014). A referential methodology for education on sustainable tourism development. *Sustainability*, 6(8), 5029–5048. https://doi.org/10.3390/su6085029.

HECoT (2012). National Qualifications Framework for Higher Education in Turkey [Online]. http://tyyc.yok.gov.tr.

Higgitt, D. (2006). Finding space for education for sustainable development in the enterprise economy. *Journal of Geography in Higher Education*, 30(2), 251–262. https://doi.org/10.1080/03098260600717331.

Ho, E. and Ang, N. (2018). Unsettling environmental management: Reflections from two environmental studies graduates in Singapore. *Environmental Education Research*, 24(2), 280–295. https://doi.org/10.1080/13504622.2016.1146661.

Hüseyinli, E. and Küçükkancabaş Esen, S. (2018). Türkiye'deki Otellerin Çevre Dostu Uygulamalarına Yönelik Müşteri Farkındalığının İncelenmesi. *Kırklareli Üniversitesi İktisadi ve İdari Bilimler Fakültesi Dergisi*, 7(3), 162–175. http://dergipark.gov.tr/klujfeas/issue/39222/457294.

Hwang, S. (2013). Placing GIS in sustainability education. *Journal of Geography in Higher Education*, 37(2), 276–291. https://doi.org/10.1080/03098265.2013.769090.

Inkson, K. and Myers, B.A. (2003). "The big OE": Self-directed travel and career development. *Career Development International*, 8(4), 170–181. https://doi.org/10.1108/13620430310482553.

Jones, P., Trier, C.J. and Richards, J.P. (2008). Embedding education for sustainable development in higher education: A case study examining common challenges and opportunities for undergraduate programmes. *International Journal of Educational Research*, 47(6), 341–350. https://doi.org/10.1016/j.ijer.2008.11.001.

Katayama, J., Örnektekin, S. and Demir, S.S. (2018). Policy into practice on sustainable development related teaching in higher education in Turkey. *Environmental Education Research*, 24(7), 1017–1030. https://doi.org/10.1080/13504622.2017.1360843.

Knudsen, D.C., Soper, A.K. and Metro-Roland, M. (2007). Commentary: Gazing, performing and reading: A landscape approach to understanding meaning in tourism theory. *Tourism Geographies*, 9(3), 227–233. https://doi.org/10.1080/14616680701422681.

Korkmaz, H. and Atay, L. (2016). Otel işletmelerinde yeşil pazarlama uygulamalarının rekabet avantajına ve işletme performansına etkisi. *Sosyal ve Beşeri Bilimler Araştırmaları Dergisi*, 38, 176–203.

Köse, İ. (2018). İklim değişikliği müzakereleri: Türkiye'nin Paris Anlaşması'nı imza süreci. *Ege Stratejik Araştırmalar Dergisi*, 9(1), 55–81. https://doi.org/10.18354/esam.329348.

Liburd, J.J. (2012). Tourism research 2.0. *Annals of Tourism Research*, 39(2), 883–907. https://doi.org/10.1016/j.annals.2011.10.006.

Marvell, A.D. and Simm, D. (2016). Unravelling the geographical palimpsest through fieldwork: Discovering a sense of place. *Geography*, 101(3), 125–136. https://www.jstor.org/stable/26546732.

McAdam, D. (1999). The value and scope of geographical information systems in tourism management. *Journal of Sustainable Tourism*, 7(1), 77–92. https://doi.org/10.1080/09669589908667327.

O'Reilly, C.C. (2006). From drifter to gap year tourist: Mainstreaming backpacker travel. *Annals of Tourism Research*, 33(4), 998–1017. https://doi.org/10.1016/j.annals.2006.04.002.

Özden, A.T. (2012). Disaster, memory and culture: Distressing attempts to develop disaster culture in Turkey. *The Newsletter, Disaster, Conflict and Social Crisis Research Network*, 13(4), 48, 9–14.

Sachs, J.D. (2012). From millennium development goals to sustainable development goals. *The Lancet*, 379(9832), 2206-2211. https://doi.org/10.1016/S0140-6736(12)60685-0.

Sadorge, J.L. (1998). The European Charter for Sustainable Tourism. *In:* Unspecified (Ed.), Natura 2000 and People: A Partnership Proceedings [EU Commission – Working Document]. pp. 27–30. Bath, UK.

Schweinsberg, S., Wearing, S.L. and McManus, P. (2013). Exploring sustainable tourism education in business schools: The honours program. *Journal of Hospitality and Tourism Management*, 20, 53–60. https://doi.org/10.1016/j.jhtm.2013.05.003.

Scott, W. and Gough, S. (2006). Sustainable development within UK higher education: Revealing tendencies and tensions. *Journal of Geography in Higher Education*, 30(2), 293–305.

Selby, D. (2006). The firm and shaky ground of education for sustainable development. *Journal of Geography in Higher Education*, 30(2), 351–365. https://doi.org/10.1080/03098260600717398.

Seremet, M. (2013). Geographic information systems (GIS) in higher education: A study of provision, pedagogy and employability in the United Kingdom and Turkey. Unpublished PhD thesis, Plymouth University: Plymouth.

Şeremet, M. (2015). Türkiye ve İngiltere yükseköğretimindeki uluslararasılaşma politikalarına karşılaştırmalı bir yaklaşım. *Journal of Higher Education & Science*, 5(1), 27–31. https://doi.org/10.5961/jhes.2015.106

Şeremet, M. (2017). Coğrafya lisans programı ve istihdama yönelik karşılaştırmalı bir değerlendirme: Türkiye ve İngiltere örnekleri. *Marmara Coğrafya Dergisi*, 36, 327–340.

Seremet, M. and Chalkley, B. (2012). Geography in Turkish higher education. *Planet*, 26(1), 59–65. https://doi.org/10.11120/plan.2012.00260059.

Şeremet, M. and Chalkley, B. (2015). Student perspectives on the teaching of geographical information systems (GIS) in geography degrees. *Journal of Geography in Higher Education*, 39(1), 18–36. https://doi.org/10.1080/03098265.2014.963795.

Şeremet, M. and Chalkley, B. (2016). Geography, GIS and employability in Turkey. *Journal of Geography in Higher Education*, 40(2), 238–253. https://doi.org/10.1080/03098265.2016.1141184.

Sheldon, P.J., Fesenmaier, D.R. and Tribe, J. (2011). The tourism education futures initiative (TEFI): Activating change in tourism education. *Journal of Teaching in Travel & Tourism*, 11(1), 2–23. https://doi.org/10.1080/15313220.2011.548728.

Simm, D. and Marvell, A. (2015). Gaining a "sense of place": Students' affective experiences of place leading to transformative learning on international fieldwork. *Journal of Geography in Higher Education*, 39(4), 595–616. https://doi.org/10.1080/03098265.2015.1084608.

Sterling, S. (2004). Higher education, sustainability, and the role of systemic learning. *In:* P.T. Corcoran and A.E.J. Wals (Eds.), Higher Education and the Challenge of Sustainability, pp. 49–70. Springer, Dordrecht.

Türkeş, M. (2001). Bonn Anlaşması ve Küresel Isınmanın Önlenmesindeki Rolü. *TMMOB Türkiye III. Enerji Sempozyumu*, 5–Aralık 2001, Ankara.

UNESCO (2003). Framework for a Draft Implementation Scheme for the Decade of Education for Sustainable Development. UNESCO, Paris.

Wade, J.A. (1999). Students as environmental change agents. *International Journal of Contemporary Hospitality Management*, 11(5), 251–255. https://doi.org/10.1108/09596119910272801.

Wallace, G. and Russell, A. (2004). Eco-cultural tourism as a means for the sustainable development of culturally marginal and environmentally sensitive regions. *Tourist Studies*, 4(3), 235–254. https://doi.org/10.1177/1468797604057326.

Weaver, D.B. and Lawton, L.J. (2007). Twenty years on: The state of contemporary ecotourism research. *Tourism Management*, 28(5), 1168–1179. https://doi.org/10.1016/j.tourman.2007.03.004.

West, P. and Carrier, J.G. (2004). Getting away from it all? Ecotourism and authenticity. *Current Anthropology*, 45(4), 483–498.

Wilson, E. and von der Heidt, T. (2013). Business as usual? Barriers to education for sustainability in the tourism curriculum. *Journal of Teaching in Travel & Tourism*, 13(2), 130–147. https://doi.org/10.1080/15313220.2013.786337.

World Economic Forum (WEF) (2018). The Travel & Tourism Competitiveness Report 2017. http://www3.weforum.org/docs/WEF_TTCR_2017_web_0401.pdf, http://reports.weforum.org/travel-and-tourism-competitiveness-report-2017/.

World Travel & Tourism Council (WTTC) (2018). Travel & Tourism Economic Impact 2019 World. https://www.wttc.org/-/media/files/reports/economic-impact-research/regions-2019/world2019.pdf.

Xu, H., Wan, X. and Fan, X. (2014). Rethinking authenticity in the implementation of China's heritage conservation: The case of Hongcun Village. *Tourism Geographies*, 16(5), 799–811. https://doi.org/10.1080/14616688.2014.963662.

Yasar, O. and Seremet, M. (2009). An evaluation of changes to the secondary school geography curriculum in Turkey in 2005. *International Research in Geographical and Environmental Education*, 18(3), 171–184. https://doi.org/10.1080/10382040903053950.

Zimmermann, F.M. (2018). Does sustainability (still) matter in tourism (geography). *Tourism Geographies*, 20(2), 333–336. https://doi.org/10.1080/14616688.2018.1434814.

Chapter 12
Integrating Environmental Protection Education in the Curriculum: A Measure to Form Awareness of Environmental Protection for the Community

Anh Tuan Hoang, Abdel Rahman Al-Tawaha, Lan Anh Vu, Van Viet Pham, Ali M. Qaisi, and Josef Křeček

1 Introduction

The integration of environmental protection education into the curriculum is essential for promoting awareness of environmental protection in societies. Sustainable Development is of great importance and Education for Sustainable Development (ESD) is critical for raising the environmental awareness of all groups of people in society, especially the young. Hence, integrating ESD and environmental education into the design of public and university education programmes is very important. Environment Protection Education helps people in general and students in particular to establish behaviours that contribute to creating a greener habitat. For a long time, people have found many ways to overpower, impose and exploit natural

A. T. Hoang (✉)
Faculty of Mechanical Engineering, Ho Chi Minh City University of Transport, Ho Chi Minh City, Vietnam

A. R. Al-Tawaha
Department of Biological Sciences, Al Hussein Bin Talal University, Ma'an, Jordan
e-mail: abdel-al-tawaha@ahu.edu.jo

Lan Anh Vu
Ho Chi Minh City University of Transport, Ho Chi Minh City, Vietnam

Van Viet Pham
Department of Science, Technology and Research Development, Vietnam Maritime University, Haiphong, Vietnam

A. M. Qaisi
Department of Pharmaceutical Sciences, School of Pharmacy, University of Jordan, Amman, Jordan
e-mail: amqaisi@ju.edu.jo

J. Křeček
Department of Hydrology, Czech Technical University in Prague, Prague, Czech Republic
e-mail: josef.krecek@fsv.cvut.cz

E. Pénzesné Kónya et al. (eds.), *Environmental Sustainability Education for a Changing World*, https://doi.org/10.1007/978-3-030-66384-1_12

resources to serve their own interests and benefits. People have exploited mineral resources, land, forests, creatures, energy, etc., to an excessive extent without caring for their resilience to serve social production and to satisfy their personal needs. While the world economy is developing strongly and gaining many achievements in science and technology, people are starting to realize that their own living environment is seriously polluted and devastated. In recent decades, negative weather phenomena and natural disasters have become increasingly intense with highly destructive power such as super typhoon Katrina (2004), Haiyan (2013), earthquakes – tsunamis (2004, 2011), floods, droughts and forest fires, which have threatened the existence of people and their livelihoods. Hence, it is time to take action to protect the environment, ecological diversity and the Earth if only to protect our own lives. Geographic science has shown that all components and landscapes of geographic shells follow certain development rules. In a unified and complete rule, all geographical components are interdependent, therefore a change of one component will lead to the change of other components. In environmental protection, we need to comply with the above rule; each positive action of individuals also brings significant effects for sustainable development and vice versa. Moreover, environmental issues and sustainable development are big and global issues that require the cooperation and efforts of all nations, communities and societies so that high efficiency can be achieved.

The slogan "Environmental education for pupils and students – the foundation of sustainable development" has great significance to the changing of perception and actions of the young generation, who needs to participate actively in environmental protection, Therefore, the role of education and training to raise environmental awareness for all classes of people in society is very important. The integration of environmental education content in the design of general and university education programmes is an essential requirement in current circumstances. Environmental education aims to help the community understand the complex nature of the system of natural and man-made environments so that people can better co-exist with their environment. Environmental education and ESD aim to equip the community with more effective environmental protection skills. The most effective method of environmental education is to equip people with knowledge on the environment in contexts that lead them to taking action to protect the environment.

Recognizing the importance of environmental education in environmental protection, many developing countries have developed specific policies and action plans to achieve remarkable achievements. Environmental education programmes, including key and extracurricular ones have been deployed at all levels of these developing countries' education systems. Environmental education consists of two main areas: education in schools and education in society. Environmental education activities in these two areas should not be separated but incorporate, support and complement each other.

2 Targeted Environmental Protection

2.1 *Basic Orientation for the Development of Environmental Protection Content (EPC) in the Curriculum*

2.1.1 Legal Basis

Since the nineteenth century, some countries have introduced environmental laws such as: The Rivers Prevention of Pollution Act in England in 1876; Coal Smoke Act in the US in 1896; Japan's Mineral Law and Japan's River Act of 1896 (Goodall 2018). In 1972, a declaration of the United Nations Conference on the Human Environment held in Stockholm addressed the challenge of preserving and enhancing the human environment showed a question on how to equip young people and adults with the morality and responsibility for protecting and improving the environment. Shortly thereafter, the United Nations Environment Program (UNEP) together with the United Nations Educational Scientific Cultural Organization (UNESCO) established an International Environmental Education Programme (IEEP). In October 1975, the IEEP organized the first International Conference on Environmental Education in Belgrade (Federal Republic of Yugoslavia). The Conference came to an agreement on the setting out of a framework decree and announcement on the objectives and guidelines for environmental education. In these documents, the goal of environmental education was to raise human awareness of the importance of the environment to human life, so providing people with knowledge on the environment to help individuals form positive attitudes to the environment, which would result in action for a better environment (Hungerford and Volk 1990). In 1987, an international conference held in Moscow on environmental education, which was organized jointly by UNESCO and the United Nations Environment Programme (UNEP), concluded on the importance of environmental education and training: "If there is no public understanding of the close relationship between environmental quality and the continuous increase in people needs, it will be difficult to reduce environmental risks locally and internationally. Human actions depend on their motivation and their motivation depends on their own perceptions and levels of knowledge." Therefore, environmental education and training is an indispensable means to help people gain full understanding of the environment (van Boeckel 2015). In Vietnam, Politburo Directive No. 36 CT/TW for June 25, 1998, called for: "Strengthening environmental protection in the period of industrialization and modernization of the country". This introduced resolutions to implement environmental protection tasks such as: "Regular environmental protection education, propagation, building of good habits and lifestyle for the Protection of Environment" and "Bringing environmental protection contents into the curriculum of all education levels in the national education system" (Bergman 2016).

2.1.2 Practical and Scientific Basis

In this twenty-first century, human beings face immense challenges. These include the over-exploitation of natural resources, exhaustion of resources, environmental pollution, and these problems have significantly affected the human utilization of natural resources. Natural resources are no longer seen as a huge "mountain" for human beings to use to serve for their essential needs of life. Once, people inconsiderately took advantage of natural resources without ever recognizing any need for "sustainable development". Today, the consequences encourage human beings to act to improve and better manage resources to ensure their sustainability (Liefländer 2015).

Education has the task of training people who not only have knowledge but also are equipped with the skills needed to meet the increasing demands of the environment and society. Training human resources to meet the requirements of the country, practically serving the socio-economic development in each field, region and locality, is the task of universities and colleges. Educating and Training produces people who will become managers, decision makers, technicians, researchers, participating in economic, political, cultural and social organizations (Russell et al. 2018). They will participate in activities that affect the living environment. Therefore, the work of education on environmental protection for university and college students with the aim of forming "experts who understand the environment" is critical for the sustainable development of the country (Ntanos et al. 2018).

2.1.3 Tasks and Basic Orientations in Environmental Protection Education

The International Association for Natural Protection (IUCN) has defined: "Environmental Protection Education is a process of recognizing values and clarifying concepts to develop the skills and perspectives needed to understand and evaluate interactive relationship between people, cultures, and the material world around us. Environmental protection education also implements an internal process of rules dealing with environmental-related issues" (Palmer 2002). According to the VIE/95/041 project in 1996, "environmental protection education is a regular process through which people are aware of their environment and gain knowledge, values, skills, experiences and determination to solve current and future environmental issues, to meet the requirements of the current generation without violating the ability to meet the needs of future generations". Environmental Education is defined as "a process that helps people gain knowledge, skills and value through formal and informal educational activities, enabling them to participate in developing an eco-sustainable society". At the Intergovernmental Conference on Environmental Education in 1977 in Georgia, UNESCO defined Environmental Education (EE) as a process of creating awareness and concern for people towards environmental issues, so that everyone is provided with knowledge, attitude, and skills to be able to develop in the future (Aikens et al. 2016). At the United Nations

International Conference on EE, held in Tbilisi in 1997, EE was determined to aim at helping individuals and communities understand the complex nature of the natural environment and artificial environment as the result of interactions between many biological, physical, chemical, economic, and social factors (Duong and de Groot 2018); so giving society knowledge, awareness of values, attitudes and practical skills empowers participation in responsible and effective ways to prevent and solve environmental issues and manage environmental quality (Wedow et al. 2018).

According to the UNESCO-UNEP Declaration in 1998, Environmental Protection Education (EPE) is not an addition to education programme, a separate part or a research topic. Instead, it should be an orientation integrated into that programme. Effective EE is the result of a reorientation and rearrangement of separate traditional disciplines and different educational experiences (natural sciences, social sciences, arts and literature, etc.), it aims to provide learners with a comprehensive environmental awareness. ESD and Environmental Protection Education (EPE) are interdisciplinary approaches that help people understand the environment with the primary purpose of care and development. It helps them develop a committed attitude, which will nurture the desire to undertake responsible actions in the environment with not only knowledge but also emotions, attitudes, skills and social actions. EE, ESD and EPC are completely inseparable from the values of practical knowledge and experience and from the ways of local and regional management and development for sustainability. Environmental education always appreciates indigenous knowledge, supports education corresponding to local environment-based learning, emphasizes the importance of both a global and local environmental perspective, and in terms of commitment and action, it is oriented towards specific localities: "Think – global, Act – Local". The information and knowledge on the environment acquired by each individual will nurture and enhance the sense of their own environmental protection responsibility, creating strong motivation and solid commitments for a fresh environment and future sustainable development. If each individual consciously contributes small but positive actions, they will help co-create a better environment. The ultimate goal of environmental education is to move towards socializing environmental issues, which means creating conscious, environmentally responsible, citizens who know how to live for the environment.

Once environmental issues have been socialized, economic benefits for the community should increase along with the effectiveness of state management while the cost burden from environmental damage should decrease. Therefore, the results of environmental studies and remedial methods in many countries around the world conclude that investing in people through environmental education is an economical and effective solution to both local and global problems. Thus, EE/EPE/ESD should be enacted across education, right from infancy to adulthood and from people working in the community to those in charge of decision-making, management, and strategic planning for socio-economic development (Fig. 12.1).

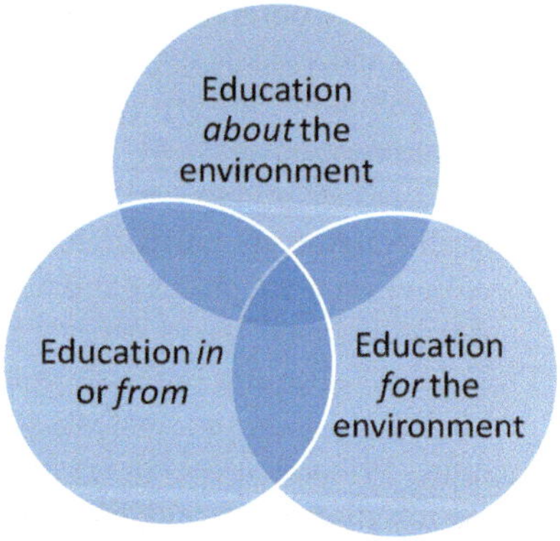

Fig. 12.1 The relationship between education and the environment (Krasny et al. 2015)

2.2 *Orientations in EPE/EE/ESD*

Environmental education aims to provide insights into the natural system and its dynamics and provide knowledge of human impacts on the environment. On this basis, environmental research and thinking skills for environmental management should be built. Environmental education aims to facilitate learning and practice in environmental practice; to build assessment and data collection skills; to foster environmental awareness and concepts; and to develop aesthetic evaluation. Environmental education aims to provide an education in the environment and for the environment; developing sensitivity to environmental responsibility and ethics; building motivation and the skills involved in improving the environment. Environmental education helps the community understand the complex nature of natural and man-made environmental systems so that people can adopt more "friendly" behaviour towards the environment. Environmental education should be applied across the whole of society, while schools, higher education and professional secondary schools, all aim to provide their students with the opportunity to:

(a) Understand the nature of environmental issues: the complexity, the multidimensional relationships, the limitations of natural resources and the load capacity of the environment, the close relationships between the environment and development, and between local, regional, national and global environments and so equip all educated people with environmental knowledge;
(b) Recognize the significance and importance of the environment as a resource for living, for working and for developing, for themselves and for their communities, nations and planet, so ensuring proper attitudes, personal values and behaviours toward environmental management and so building correct conceptions of

responsibility. This goal is oriented to attitude building and to learning to treat the environment in a friendly and supportive way, conscious of its needs and limitations;

(c) Possess the knowledge, skills and methods of action needed to improve their ability to choose suitable lifestyles for the rational and wise use of natural resources, so that they can participate effectively in preventing and solving specific environmental issues where they live and work. This is a feasible goal.

2.3 *Aims of Environmental Education*

Environmental education should address the following five general aims. It should:

1. Be integrated and comprehensive: Effective environmental education is a comprehensive integrated education designed to equip students with environmental knowledge through each individual subject and specific programme. This education is founded on integrated methods and exploring the relevant aspects of the teaching content of each suitable subject (e.g. Cihangir and Şeremet 2020).
2. Be appropriate: Environmental education is the provision of appropriate environmental information and environmental protection measures in ways suitable for each educational level and for the cognitive ability of each group. The content of environmental education needs to be pro-environmental action; it should help people be aware of their responsibilities and build practical skills to protect the environment.
3. Engage with synthesis and diversity: In social life, natural factors and social factors always have interactions and many directly affect the daily life of individuals and residential communities; therefore, the content of environmental education should include not only the content of the natural environment but also the social or human in all its diversity.
4. Evoke practical action: Environmental Protection Education should not only help students have greater awareness and essential knowledge to protect the environment, but also address the principles, rules and tools needed to know what to do for the surrounding environment. EPE/EE/ESD must be both practical and encourage necessary action. This education should enable learners to apply their knowledge to the processes of solving environmental problems (Merenlender et al. 2016).
5. Be collaborative: EPE/EE/ESD should teach students how to behave and act collectively in support of the environment. Hence, it is necessary to make use of cooperative methods between teachers and learners and between schools and society in the process of education (Fig. 12.2).

Fig. 12.2 Citizenship and environmental protection (Duong and de Groot 2018)

3 Objectives in Designing Programme Content that Integrates Environmental Protection Education into the Curriculum

3.1 Objectives

Through the integration of EPE content in the curriculum, students and pupils can be provided with knowledge, skills and methods of environmental action, specifically:

1. Knowledge: Basic knowledge of environmental science and the real situation of natural resources and human activities leading to natural resources depletion and degradation.
2. Skills: Skills needed to identify behaviours that violate the environment and to take specific measures that contribute to environmental protection and remediation.
3. Attitude: Value-driven attitudes and ethics that enable students to be aware of the current environmental situation in order to adopt proper behaviour and promote environmental protection activities.

3.2 *Tasks*

Regarding knowledge, the main tasks are: to providing some basic knowledge of the environment such as natural environment, social environment, environmental pollution, the impact of environmental pollution on human life and the ways, measures and forms of developing environmental education content, not least for preschool children.

Regarding skills: the main tasks are to: build the action skills and behaviours appropriate to the living environment at tertiary level and build teaching organization skills and methods of integrating environmental education content – not least at preschool level.

Regarding attitude: the main tasks are to help students realize the significance and importance of environmental issues to themselves and to their community at national and international level, which will help them build positive and constructive attitudes towards environmental issues, a sense of personal responsibility and a wish to engage with voluntary participation in environmental protection activities (Fig. 12.3).

Fig. 12.3 Education to protect the environment (Goodall 2018)

4 Some Ways and Means to Integrate Environmental Education in the Curriculum

4.1 Environmental Protection Education in Preschools

Preschool education is part of the national education system, which is an important foundation for the initial formation of personality and future conduct. The period of 0–6 years is a very important time in every person's life. During this period, children develop very quickly in terms of physical, cognitive and emotional aspects. Social relationships, personality traits, qualities and general abilities are usually formed in this age group (Ntanos et al. 2018).

Teaching children from a young age how to protect the environment is a very good thing. Environmental education for preschool children is to provide basic knowledge of the environment in accordance with the child's cognitive ability, so creating positive attitudes and behaviours towards the environment around them. Determining that environmental protection education is an important task, preschools can design special programmes suitable for each stage and each age group. This is a mission that needs specific strategies and plans to develop a child's personality comprehensively. EPE can be integrated into daily activities so that children can learn, play, and acquire knowledge in the most effective way (Russell et al. 2018).

Teaching children basic concepts includes helping them understand basic concepts such as: clean environment, dirty environment, why the air is polluted, what harm environmental pollution causes, why hands need washing before eating and after going to the toilet, why it is necessary to protect and plant trees, etc. (Szczytko et al. 2018) and encouraging children to look for information on environment through contests such as finding out about their favourite topics with their parents and giving presentations in their class. This activity is suitable for 5-year old classes. Through such activity, children may become very excited with what they learn and explore and remember them longer. Some children have many great ideas to protect the environment.

4.2 Carrying Out the Most Practical Actions at the Preschool

Guiding children to form habits such as cleaning tables and chairs, arranging toys and utensils neatly, folding the blanket after waking up and arranging their lockers neatly. Littering or leaving the class untidy should be absolutely prohibited. There will be a group of friends who observe these activities so that each team of children can compete with other teams to see which one is cleaner and neater. Public labour activities can be carried out at weekends. The children will walk around the classrooms and the school to collect garbage, clean the toy racks, rearrange items, etc. Teachers will write warnings with funny pictures like "Saving water" on hand washers and common used water bottles, "Please give me garbage" on the garbage bin,

"Wash your hands after going to the toilet" in the toilet, Children will gradually memorize the notices and voluntarily follow them. The seeding and planting lessons are the most exciting hours for children. Children can grow a young plant with their parents and then bring them to class or they can plant trees in the school garden with their teachers and friends. This activity helps children understand the practical meaning of planting trees to create a beautiful green environment and the need to protect the trees around them (Pande and Pande 2020) (Fig. 12.4).

Further important activities include recycling in the classroom: teachers will find toys and utensils made from recycled materials on the internet, books and newspapers then with the children, they make their own versions in class. Recyclables can be made using materials like beer cans, eggshells, newspaper, toilet paper core, cardboard, straws, etc. There are children who are very quick and can perform this work with parents at home and bring pretty products to class (Varela-Candamio et al. 2018), as well as make 'propaganda' to parents. Outside class hours, teachers believe that children can learn a lot of environmental protection knowledge from their families. Kindergarten teachers may propagate the most simple and effective measures to take when they are at home. The children become "supervisors" reminding parents to save water, turn off the electricity after use, sort proper waste, or collect unnecessary supplies to donate to charitable organizations. When children feel that they have become important people, they will do things very well. As a result, lessons on environmental protection will promote effects and meaning into the community.

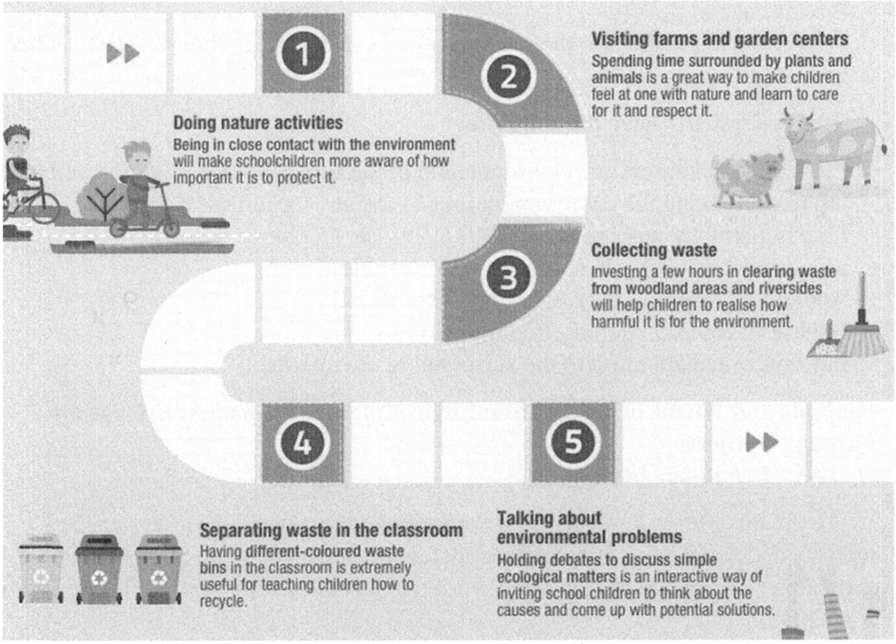

Fig. 12.4 Five ways to teaching children about the environment (Ardoin et al. 2018)

5 Environmental Protection Education at Primary School Level

Primary school is the basic level and an important platform for educating and training children to become good citizens for the country. The important purpose of environmental protection education is inform but also inculcate habits and behaviours that are civilized and friendly to the environment. If at this level of education, the children have not yet formed a love for the nature, a harmony with nature, a care for the world around them, a habit of tidiness and hygiene, then it will be difficult for them to develop these in their later education. Therefore, the content and method of environmental education in primary schools is crucial to the formation of those qualities. Developing content and programme of environmental protection education awareness for primary school children in the direction of enhancing the integration of some closely related contents of the subjects and activities outside class should avoid repetition and, more importantly, create a synergy between awareness of causes and solutions to the current environmental problems of life (Mustam and Daniel 2018; Otto and Pensini 2017).

The objectives of environmental protection education at primary school level are to make children understand and learn about:

- Environmental components of soil, water, air, light, animal and plant, and the relationships between them.
- The relationship between people and environmental components.
- Environmental pollution.
- Protection measures for their surrounding environment (house, class, school, hamlet, village, street).

Children initially have the ability to:

- Participate in appropriate environmental protection activities (planting and tending trees; making the environment green – clean – beautiful.
- Live in harmony with nature.
- Live tidily and neatly in sharing and cooperation.
- Appreciate nature, family, school and homeland.
- Behave in environmentally friendly ways.
- Pay concerned attention to the surrounding environment.

Methods and Forms of Integrating Environmental Protection Education Through Subjects

(a) *Level 1 (Full integration)*

For the integration lessons of environmental education at this level, teachers help students understand and fully and deeply feel the lesson content. Such lessons allow environmental protection content to work for the children through each topic (Cruz et al. 2018).

(b) *Level 2 (Partial integration)*

When teaching integrated lessons at this level, teachers need to study the content of the lesson carefully, determine what content of the environmental protection education is integrated into the lesson, what content and what activities of the teaching process is the environmental protection education integrated into, and what teaching aids and teaching facilities need preparing (de Andrade et al. 2018). When organizing teaching, teachers employ normal teaching activities in accordance with the organization and teaching methods of the subject. In the process of organizing teaching activities, teachers aim to help children understand and fully connect with the content of the lesson related to environmental protection, which contributes to educating children naturally in environmental protection awareness. Teachers should gently and appropriately integrate these contents to achieve the objectives of the lesson as required by the subject (Ntanos et al. 2018).

(c) *Level 3 (Relevant integration)*

When preparing lessons, teachers should have a sense of integration, prepare suggestive and related issues in order to educate students on environment protection and living and learning skills in a sustainable development environment. When organizing teaching, teachers organize normal teaching activities, in accordance with the organization and teaching methods of the subject. In the process of organizing teaching activities, teachers organize, guide students to relate to and expand on environment protection education in a natural, harmonious and proper way, avoiding rambling and putting students under pressure, not suitable for specific subjects (de Andrade et al. 2018).

The applicable methods with either the whole class or a group of children are: discussion methods; methods of observation; gaming methods; methods of inquiry and investigation. This EE/ESD may be integrated into class periods, outside the classroom as in the local environment, in school through the practice of cleaning classrooms and keeping schools and classrooms clean and beautiful (Fig. 12.5).

6 Environmental Education at University Level

Here, the content should include:

1. Basic concepts of environment: environment, environmental pollution, natural phenomena such as weather, climate change, acid rain, greenhouse effect etc.,
2. Environmental issues: environmental status, environmental pollution, causes of environmental pollution, mineral resources...
3. Measures: ways to protect the environment, environmental protection awareness, mitigating actions that adversely affect the environment, some ways to improve environmental quality...
4. Special topics including aspects of environmental pollution, waste disposal, deforestation, population growth and human needs.

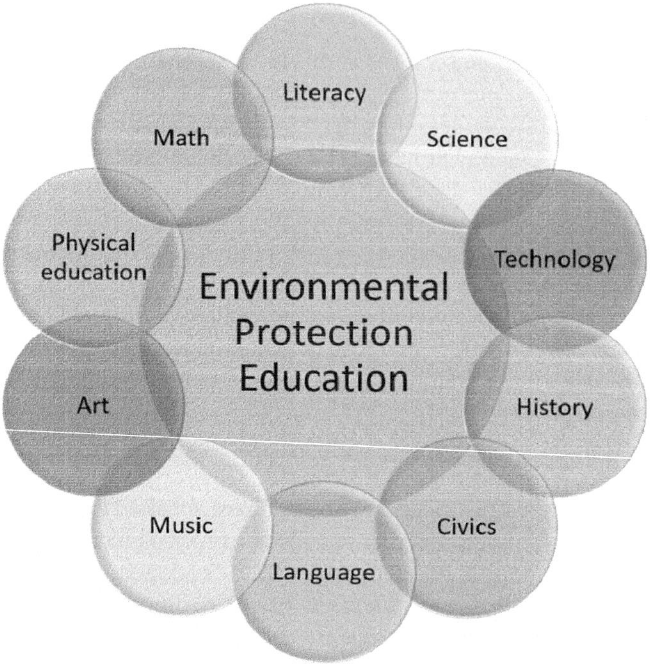

Fig. 12.5 Integrating environmental protection education into high school subjects (West 2015)

Implementation may be organized across the curriculum by:

- Lectures, exercises and practical experience of learners: Guiding and helping students to design images or videos describing the environment according to each topic, having students do the assessment tests and give presentations at the class level (de Andrade et al. 2018).
- Sightseeing and field surveys: Organizing visits to some specific places, either inside or outside the school to help students learn how to assess and relate the knowledge to actual situations.
- Problem solving: Raising a problem and asking students to apply their own understanding to solve environmental problems from a personal perspective.

6.1 Specific Integration Measures in the University Curriculum

Determining the priority of each subject helps build the content and methods of environmental education integration according to the characteristics of each subject. The first measure: In schools where students have a choice of courses, it is possible to develop a course programme on 'Environment, Society and Sustainability' for students to choose. In this module, it would be necessary to develop a curriculum of about 60% theory and 40% practice. This would help equip students with both scientific knowledge and skills to build environmental protection activities.

In addition, universities, as a whole, could identify themes that integrate the content of environment education and training at different levels as a priority. Based on the curriculum of each discipline, universities can build a system of subjects that include and integrate environment education (Alföldi 2020).

The process of integrating environmental protection education in the curriculum involves:

Firstly: Before implementing the integration of environmental protection education contents into the curriculum, it is necessary to establish a specialized department to manage all the integrated contents, including which subjects to integrate and which knowledge should be integrated. The aim is to avoid overlap and repetition and to promote the integration of environmental protection and sustainability education content (Pedrini et al. 2016).

Secondly: Adequate attention should be paid to whether the curriculum has a Human Environment or Environmental Education component or not to determine the content of knowledge and organize the integration process effectively.

Thirdly: Too much attention should not be paid to the knowledge of environmental science but it is necessary to strengthen environmental education awareness among students.

Fourthly: Strengthening the active learning of students in the integration of environment education and training helps students self-identify problems through their own experience and grants them the opportunity to make comparisons and to evaluate theory and practice.

7 Conclusions

Environment Protection Education (EE/ESD/EPE) helps people in general and students in particular become aware of their role and responsibilities to the environment and also helps set up specific behaviours that contribute to promoting the campaign "For a Greener country", which becomes more than a guiding document or chanted slogan, but something creating cleaner streets, more decent and civilized cities, and a better habitat in general, where even the behaviour between people becomes more sincere and harmonious. In this way, the meaning of this new campaign becomes practically affective leaving deep and memorable impressions in each person's mind (Dania and Sunday 2017).

Universities and colleges are starting to build a curriculum of Environmental Education that helps students form inclinations and habits to protect the environment. They should equip their students with basic knowledge about the environment, environmental protection and sustainability. Their students should be given favourable conditions to design or organize their personal integrations of environmental knowledge and environmental protection strategies from each of their special subjects of the curriculum. Appropriate content should be integrated into all subjects in the curriculum, if only to reduce the work load for the Environmental Education departments. If there is not enough time to integrate theoretical knowledge fully, it is recommended to strengthen group and practice activities with subject matter related to environmental protection. Union organizations in the school should organize practical activities to protect the environment ahead of slogans and exhortation. Groups or Clubs "For a Beautiful Life" or "For a More Beautiful City" should be established and empowered to undertake practical and effective tasks.

References

Aikens, K., McKenzie, M. and Vaughter, P. (2016). Environmental and sustainability education policy research: A systematic review of methodological and thematic trends. *Environmental Education Research*, 22(3), 333–359.

Alföldi, Z.P. (2020). An integrated model of sustainability education. *In:* Pénzesné Kónya, E., Haigh, M. and Křeček, J. (Eds). Environmental Sustainability Education for a Changing World: Inspiration for and by Practitioners. Springer, Cham, Switzerland & Capital Publishing Company, New Delhi, pp. 223–244.

Ardoin, N.M., Bowers, A.W., Roth, N.W. and Holthuis, N. (2018). Environmental education and K-12 student outcomes: A review and analysis of research. *The Journal of Environmental Education*, 49(1), 1–17.

Bergman, B.G. (2016). Assessing impacts of locally designed environmental education projects on students' environmental attitudes, awareness, and intention to act. *Environmental Education Research*, 22(4), 480–503.

Cihangir, E. and Şeremet, M. (2020). Education for Sustainable Development (ESD) in Tourism-oriented vocational degree programmes in Turkey. *In:* Pénzesné Kónya, E., Haigh, M. and Křeček, J. (Eds). *Environmental Sustainability Education for a Changing World: Inspiration for and by Practitioners*. Springer, Cham, Switzerland and Capital Publishing Company, New Delhi.

Cruz, A.R., Selby, S.T. and Durham, W.H. (2018). Place-based education for environmental behavior: A 'funds of knowledge' and social capital approach. *Environmental Education Research*, 24(5), 627–647.

Dania, P. and Sunday, O. (2017). University Social Studies Students' Knowledge of the Components of Environmental Education and Their Participation in Environmental Protection. *Advances in Social Sciences Research Journal*, 4(2), 59–67.

de Andrade, J.B.S.O., Garcia, J., de Andrade Lima, M., Barbosa, S.B., Heerdt, M.L. and Berchin, I.I. (2018). A proposal of a Balanced Scorecard for an environmental education program at universities. *Journal of Cleaner Production*, 172, 1674–1690.

Duong, N.T. and de Groot, W.T. (2018). Distributional risk in PES: Exploring the concept in the Payment for Environmental Forest Services program, Vietnam. *Forest Policy and Economics*, 92, 22–32.

Goodall, S. (2018). Developing Environmental Education in the Curriculum. Routledge.

Hungerford, H.R. and Volk, T.L. (1990). Changing learner behavior through environmental education. *The Journal of Environmental Education*, 21(3), 8–21.

Krasny, M.E., Kalbacker, L., Stedman, R.C. and Russ, A. (2015). Measuring social capital among youth: Applications in environmental education. *Environmental Education Research*, 21(1), 1–23.

Liefländer, A.K. (2015). Effectiveness of environmental education on water: Connectedness to nature, environmental attitudes and environmental knowledge. *Environmental Education Research*, 21(1), 145–146.

Merenlender, A.M., Crall, A.W., Drill, S., Prysby, M. and Ballard, H. (2016). Evaluating environmental education, citizen science, and stewardship through naturalist programs. *Conservation Biology*, 30(6), 1255–1265.

Mustam, B. and Daniel, E.S. (2018). Informal and Formal Environmental Education Infusion: Actions of Malaysian Teachers and Parents among Students in a Polluted Area. *MOJES: Malaysian Online Journal of Educational Sciences*, 4(1), 9–20.

Ntanos, S., Kyriakopoulos, G., Arabatzis, G., Palios, V. and Chalikias, M. (2018). Environmental Behavior of Secondary Education Students: A Case Study at Central Greece. *Sustainability*, 10(5), 1663–1685.

Otto, S. and Pensini, P. (2017). Nature-based environmental education of children: Environmental knowledge and connectedness to nature, together, are related to ecological behaviour. *Global Environmental Change*, 47, 88–94.

Palmer, J.A. (2002). Environmental Education in the 21st Century: Theory, Practice, Progress and Promise. Routledge.

Pande, A. and Pande, L. (2020). Community learning and resilience in the Himalayas: Uttarakhand, India. *In:* Pénzesné Kónya, E., Haigh, M. and Křeček, J. (Eds). Environmental Sustainability Education for a Changing World: Inspiration for and by Practitioners. Springer, Cham, Switzerland & Capital Publishing Company, New Delhi, pp. 45–62.

Pedrini, A.d.G., Brotto, D., Pimentel, D.d.S., Behrends, E. and de Moraes, A. (2016). Transformative and Emancipatory Environmental Education by Marine Ecotourism in the marine environmental protection area of Armação dos Búzios (RJ, Brazil). *Revista Brasileira de Ecoturismo*, 9(3), 445–470.

Russell, C., Gough, A. and Whitehouse, H. (2018). Gender and Environmental Education in the time of# MeToo. Taylor & Francis.

Szczytko, R., Carrier, S. and Stevenson, K.T. (2018). Impacts of outdoor environmental education on teacher reports of attention, behavior, and learning outcomes for students with emotional, cognitive, and behavioral disabilities. *Frontiers in Education*, 19th June 2018, 46. doi: https://doi.org/10.3389/feduc.2018.00046. https://www.frontiersin.org/articles/10.3389/feduc.2018.00046/full.

van Boeckel, J. (2015). At the heart of art and earth: An exploration of practices in arts-based environmental education. *Environmental Education Research*, 21(5), 801–802.

Varela-Candamio, L., Novo-Corti, I. and García-Álvarez, M.T. (2018). The importance of environmental education in the determinants of green behavior: A meta-analysis approach. *Journal of Cleaner Production*, 170, 1565–1578.

Wedow, R., Zacher, M., Huibregtse, B.M., Mullan Harris, K., Domingue, B.W. and Boardman, J.D. (2018). Education, smoking, and cohort change: Forwarding a multidimensional theory of the environmental moderation of genetic effects. *American Sociological Review*, 83(4), 802–832.

West, S.E. (2015). Understanding participant and practitioner outcomes of environmental education. *Environmental Education Research*, 21(1), 45–60.

Chapter 13
An Integrated Model of Sustainability Education

Zoltán P. Alföldi

1 Introduction

For thousands of years, people have been living close to, or actually in, nature spending most of their times with plants and animals—that is, in real life. It is assessed that Csángó people living in the Gyimes region of Romania are still spending between 200 and 210 days in a year in nature (Babai and Molnár 2013, personal communication). However, today, as a result of radical changes in lifestyle related chiefly to increased industrialization and urbanization, most people are living lives more separated from nature and students no longer spend their childhood in natural settings. This trend has such significant multiple effects on the human psyche and behaviour, that Richard Louv (2005) introduced the term 'Nature Deficit Disorder' (NDD), as a way to describe the human cost of alienation from nature, whose consequences include a diminished use of the senses, attention difficulties and higher rates of physical and emotional illnesses.

The early quotation from John Muir from 1890, that "Most people are 'on' the world, not in it—have no conscious sympathy or relationship to anything about them…" (Mossberg 2005, p. 193) holds especially true now in industrial societies. In the real world, however, "we *are* related, indispensable to each other for our most basic life requirements" (Macauley 1996, p. 127). Finding the proper ways for human flourishing within the complex, interrelated, multilevel ecological, social and economic system is of fundamental significance and seems to be very difficult to realize. To this end, sustainability is a goal to accomplish with theoretical knowledge and proper action (York and Becker 2012). Environmental education programmes have very significant roles in helping the broad public realize that pro-environmental behaviours are directly connected to prosocial behaviours and

Z. P. Alföldi (✉)
Department of Plant Sciences and Biotechnology, Georgikon Faculty, University of Pannonia, Keszthely, Hungary
e-mail: alfoldi.zoltan@georgikon.hu

E. Pénzesné Kónya et al. (eds.), *Environmental Sustainability Education for a Changing World*, https://doi.org/10.1007/978-3-030-66384-1_13

serve human interests (Neaman et al. 2018). There is an increasing global trend in universities to incorporate the integrative view of sustainability into various forms of creative and effective sustainability education programmes, as they have a key leadership role in finding reasonable solutions for compelling eco-social issues and so shaping more aware societies (Ralph and Stubbs 2014; Filho et al. 2015). These establish the claim for applying dynamic and effective methods of pedagogy in environmental education in which students and teachers work together to find solutions for problems of relevance to their communities and to accomplish 'complex-term' sustainability (Fien and Tilbury 1996).

Sustainability refers to the entire web of relationships and interactions in the world. It is both an individual and a community practice (Capra 1997). The fact that ecological sustainability is a property of a web of relationships means that in order to understand it properly, that is to become ecologically literate, we need to learn how to think in terms of relationships, in terms of interconnections, patterns and contexts. In science, this type of thinking is known as systems thinking. Systems thinking highlights the importance of relationships. "All the essential properties of a living system depend on the relationships among the system's components. Systems thinking means thinking in terms of relationships" (Capra 2011, p. 52). It is crucial for understanding ecology, because ecology is the science of relationships among the uncountable members of the Earth's biota. Sustainability means engaging with the enormous variability of dynamic processes that promote various life forms (Capra 1997, 1999). When we become ecologically literate, we can easily realize that ignoring and interfering with ecological patterns and processes are the fundamental causes of many of our current world problems (Capra 2011).

In an interview, Capra defines an ecologically sustainable society as "a society that is designed in such a way that its ways of life, businesses, economy, physical structures, technologies and social institutions do not interfere with nature's inherent ability to sustain life" (Pisani 2007, p. 24). Learning how not to interfere with this ability of nature requires that we possess that system of knowledge called ecoliteracy. This includes the basic principles of ecology, and the understanding that these constitute the fundamental rules of sustainability. As Capra states, "If we recognize certain laws of nature and recognize that disregarding these laws we will harm ourselves, then we had better take it into account" (Pisani 2007, p. 17). Capra continues: "One key principle is the network as the fundamental organizing principle of ecology," and not only "of ecosystems, but of living systems in general" (Pisani 2007, p. 18).

The human-nature relationship and its worldviews need to "encompass interconnectedness and the interrelatedness of environmental and social issues" (Jordan and Kristjánsson 2017, p. 1207). It is of utmost importance for students to be familiar with environmentally significant ecological concepts and processes and to understand the interrelationships that exist within and between these (Hungerford and Volk 1990). The primary goal and mission of environmental education is to advance ecological literacy. The most characteristic features of ecoliteracy, as it was used by Capra (1997), building on David Orr's work (Orr 1992, 2000), are its focus on human and environmental sustainability and the emphasis placed on its spiritual component.

Hence, the curriculum of a Human Ecology course should be based on system thinking, which incorporates the recognition of the various biophysical and socio-economical components in particular environmental context, and reveal their inter-relations. Historical and recent lessons help reveal and understand complex ecological processes, in the service of proper, wise decision-making.

The goal is to educate people in an ecological-holistic way of thinking and to reorient their actions henceforth within a new environmental paradigm. In addition, our firm belief is that the task should be to reveal and learn intercultural wisdoms, to incorporate the relevant issues of traditional ecological knowledge of local societies and novel scientific results and to live in co-operative, knowledge- and ethic-based societies. These could be the foundations for the surviving in the midst of the various crisis of the Anthropocene.

2 Burdens for Sustainability Education

Our relationship with the environment in which we exist is shaped by our perspectives, assumptions, habitual attitudes, and these in turn are shaped by our heritage and by the cultural values we have imbibed. To live prosperously on the 'land' (in *Aldo Leopold*'s sense) and keep it healthy and sustainable, we need to develop ecological consciousness, with its key issues of interconnection and interdependence. Education, raising public awareness, is fundamental to achieve a sustainable future. Sustainability education programmes at all levels are of key importance for our multi-species communities in the human-dominated epoch of the Anthropocene. However, there are significant challenges in this field, which are experienced as obstacles to and burdens upon effective sustainability teaching. According to our experience, the most important of these are:

- **'Knowledge-Boxes':** The fragmented and separated fields of knowledge disciplines have the consequences of producing not systematical but disjointed knowledge and partial understanding.
- **Information chaos:** There is so much confusing information and misinformation; we become overloaded and overwhelmed and often miss the key messages, resulting in uncertainty.
- **Inter-generational sloth or excessive comfort:** We have known lots of (inconvenient) truths about our world and its growing environmnetal crises for a long time—so why are we so inactive and failing to address them? Why do so many not care that our "house is on fire" as Greta Thunberg argues (Thunberg 2019)?
- **The 'Captain Planet Approach':** 'A super-hero will eventually save us' attitude, or excessive technical optimism fuels the view of the future.

Nature was considered differently in the previous spiritual and ideological constructs linked to particular historical, social and economic conditions. It was seen as "a terrestrial paradise; divinely ordained human stewardship over nature; nature's death and plunder; nature as a primordial wilderness, nature as nurturing mother;

and nature as brute matter, to name a few", as it was cited by referring to Adler (2006). Today, we need to have proper policy frameworks, theoretical guidelines and practical tools for efficient environmental education to achieve sustainability goals.

3 An Integrated Model for Sustainability Education

At the Georgikon Faculty of the University of Pannonia, Keszthely, Hungary, the response has taken the form of a 2017, Hungarian TEMPUS award-winning, integrated (both in structural and functional terms) model of sustainability education invented by a team of university professors and students, and developed for the last 15 years. This complex system consists of several component parts, including:

- A curriculum that includes courses on: **Human Ecology** (focusing on the interrelations of the different 'spheres', as shown in Fig. 13.1), **Agricultural and Environmental Bioethics** (with special emphasis on cross-cultural values and virtues), and **Environmental Education** (both theoretical and practical approaches), that aims to develop efficient and useful competencies for students and introduce them on how to live in co-operative, knowledge-, and ethic-based societies.
- The development of **environmental** and **ecological literacy** among students as the major goal for sustainability education. Case studies cover lessons from the past with relevant and useful consequences.
- The significance of **environmental virtue ethics**, considering the particular importance of virtue as a positive character trait. The relevance of virtue ethics developed within different cultures, world views and religions have already proved important in sustainability education.
- The pedagogy of Joseph Kentenich (**'Schoenstatt-pedagogy'**), a system built on deep psychological and spiritual foundations.
- **Study-tours** and the **'Club for Land and Humans'** which is a kind of an extra-curricular student club run by students who, working closely with their profes-

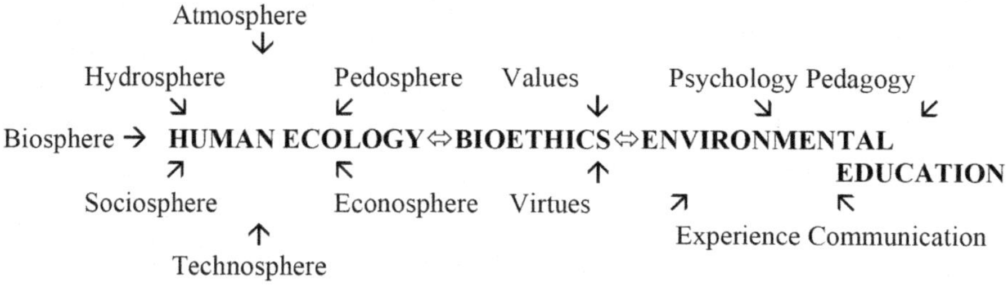

Fig. 13.1 Outline of the integrated model of sustainability developed at the Georgikon

sors at the Georgikon, provide their peers with relevant practical experiences, particularly sharing the traditional ecological knowledge of local societies.
- An **e-learning** curriculum for adult education, and a professional vocabulary, as well as an informative study-aid about business development for specialists working in the field of environmental protection.

These integrated units, as presented in Fig. 13.1, are used at the different levels (BSc and MSc) of sustainability education, in both Hungarian and English (for foreign students). This programme is integrated in the system of sustainability education at the University of Pannonia, and already has been proven to be very effective and useful for this complex field and for demonstrating its practical relevance.

4 Fundamentals of the Integrated Model

As Orr pointed out: "all education is environmental education" (Orr 1994, p. 12). The experience of nature is rooted in social and cultural experience. As personal identity emerges in a social context, pro-environmental action is facilitated when peoples perceive nature as a moral entity and not just a resource; when the social environment maintains the feelings and realizations of connectedness with nature, and when social contexts support pro-environmental identities (Clayton and Opotow 2003, pp. 1–24). Significant life experiences (Chawla 2006) can yield the realization of awe in nature that fosters the passionate love for it, and thus a "deeper, richer sense of nature" may become integrated into environmental consciousness (Swan 2010, p. 5).

We are born with the innate ability and sensitivity to perceive and absorb the endless wonders of life and beauties of its inconceivably rich variety, but the intention and education needed to maintain and develop it are also necessary to support such deeply-rooted love of the natural world. If one can keep environmental sensitivity (the sense of wonder) from his or her childhood and, later on, in adulthood allow the possibilities to his or her experiences in nature, this can serve as a source of physical and spiritual strength (Carson 1998; Kato 2015). As Rachel Carson puts it, "it is more important to pave the way for the child to want to know than to put him on a diet of facts" (Carson 1998, p. 45).

It is proven that our brain uses, typically, unconscious cognitive structures called 'frames' in the process of thinking (Lakoff 2010). Many frame-circuits have direct connections to the brain's emotional regions. Mental frames are working in systems and we use them in relation to all of our knowledge by neural activation. Therefore, effective environmental education strategies, have to change old—and harmful, in their consequences—frames. Here, the use of proper (new) language in communication is of primary importance because this allows new, sustainable ideas and information to spread through the broad population (Lakoff 2010).

As the Belgrade Charter of the UNESCO-UNEP (1976) states, "The goal of environmental education is to develop a world population that is aware of and concerned about, the total environment and its associated problems, and which has the knowledge, attitudes, skills, motivation, and commitment to work individually and collectively toward solutions of current problems and the prevention of new ones" (UNESCO-UNEP 1976; cited in Athman and Monroe 2001, p. 38). In 1977, the first intergovernmental conference on environmental education was organized in Tbilisi, Georgia. The Tbilisi Declaration outlined the following five categories of environmental education objectives (UNESCO-UNEP 1978; cited in Athman and Monroe 2001, p. 38):

- **Awareness** and sensitivity to the complex environment and its problems;
- **Knowledge:** from a variety of experiences in, and acquire a basic understanding of, the environment and its associated problems;
- **Attitudes:** a set of values and feelings of concern that are positive for environment and motivation for actively participating in environmental improvement and protection;
- **Skills** acquired for identifying and solving environmental problems; and
- **Participation** or **behaviour** that encourages citizens to be actively involved at all levels in working toward the resolution of environmental problems (UNESCO-UNEP 1989).

Based on this declaration and other research, Hungerford and Volk (1990) identified the effective components of a successful environmental education programme. Environmental educators should aim to improve the environmental literacy of the learners, foster their enthusiasm, increase positive attitudes toward the environment, competency in environmental action skills, and have the role of "preparing students to become critical thinkers, informed decision-makers and able communicators—a role that exceeds far beyond presenting information" (Athman and Monroe 2001, p. 37).

Environmental literacy depends on a personal commitment and motivation to improve environmental quality and, consequently, quality of life. This commitment and motivation often begins with an awareness of one's local, immediate surroundings. For motivation, personal experiences are of particularly great relevance in helping students develop a sense of empowerment and recognize their personal responsibilities (Athman and Monroe 2001).

An effective environmental education programme has the following major features according to Staniskis and Stasiskiene (2006):

- It considers the environment in an integrated way, that is, in its full ecological, historical, socio-cultural, economic, political and technological cross-disciplinarity.
- It responds to the challenge of the complex and dynamic nature of environmental, social and economical sustainability.
- It focuses on the multi-level interactions between the natural-physical and the socio-economic environments.
- It harmonizes knowledge transformation, skill development, capacity building and attitude formation in order to achieve commitment to pro-environmental action.

- It recognizes the value of traditional ecological knowledge and its practical applications in reaching the realistic goals of sustainability.
- It promotes education learning by practice, which means that study tours and field trips are important constituents of the curriculum.
- It considers both the local and global environments as interrelated systems, so it includes local, national and international perspectives on environmental issues.
- It focuses on recent and future perspectives on environmental conditions.
- It emphasizes participation in preventing and solving environmental problems.
- It involves ethical principles and argumentation in discussing moral dilemmas.

5 Theoretical Models of Environmental Behaviour

A variety of behavioural change models have been proposed in recent decades (Hines et al. 1987; Hungerford and Volk 1990; Kollmuss and Agyeman 2002; Akintunde 2017). Some are introduced here briefly. In the early days, environmental behavioural models were **simple linear models,** which emphasized that environmentally favourable attitudes are created through increasing knowledge, so leading to responsible environmental actions (Fig. 13.2).

The traditional environmental education model postulated that providing a person with knowledge about a particular environmental crisis would, in the end, result in that person adopting the appropriate behaviours. However, although information and knowledge are necessary, simply providing—even the most accurate—facts may not lead to significant changes in behaviour (Hines et al. 1987). Boyes and Stanisstreet (2012) argue that there is no strong link between a person's general environmental attitudes and knowledge, and their willingness to undertake pro-environmental actions. As Frasz (1993, p. 269) states, "we cannot be sure that knowledge of nature will produce caring for nature". Consequently, when empirical evidence suggested that environmentally responsible behaviour (ERB) results from the interplay of multiple variables, and it was realized that these variables interact in different degrees in influencing ERB, the legitimacy of the simplistic linear human behavioural change model was challenged and replaced by more interactive models (Hungerford and Volk 1990; Akintunde 2017).

The **theory of environmentally responsible behaviour** (ERB), proposed by Hines et al. (1987), stresses the importance of multiple variables including: intention to act, locus of control (LOC) (an internalized sense of personal control over the events in one's own life), attitudes, sense of personal responsibility, and knowledge. The model is based on the assumption that having the intention to act is a

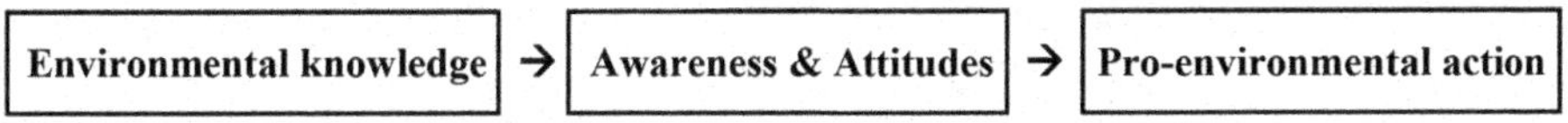

Fig. 13.2 The schematic representation of early models of pro-environmental behaviour. (Adapted from: Kollmuss and Agyeman 2002; Akintunde 2017)

major factor influencing ERB, and that 'internal control' has a special impact on the intention of acting, which affects the ERB of the person.

The ERB model also highlights the relationship between the LOC, the attitudes of the individuals, and their intention to act. Inner controls directly affect a person's attitudes, which can lead to a higher intention of acting. This suggests that knowledge itself is insufficient to generate responsible pro-environmental action but rather that this depends upon the interactions of many additional factors (Hines et al. 1987). This underlines the importance of developing environmentaly responsible attitudes, which can be translated to action if personal circumstances permit. Thus both internal and external factors control both attitudes and intentions of acting and may form a base from which learnt predispositions for pro-environmental behaviour are converted into actions (Hines et al. 1987; Akintunde 2017).

The ERB can reasonably be defined by its direct or indirect impacts (impact-oriented definition), that is, how much it changes the exploitation of resources taken from the environment or affects the structure and dynamics of ecosystems. Alternatively, ERB can be defined from the actor's intention to change (benefit) the environment (intent-oriented definition) (Stern 2000). As a primary goal of environmental education, students must be educated to be able to make the direct connection between the importance of their ERB and their perceived direct or indirect benefits, either in the short-term or longer term, and so habituate ERBs into their everyday lives (Ungerer 2015).

The **Environmental Citizenship Model** was proposed by Hungerford and Volk (1990). In this model, the variables that influence whether a person takes action were grouped into three stages of educational involvement ranging from entry level (including general sensitivity to and knowledge of the environment), through ownership level (characterised by in-depth knowledge, personal commitment, and resolve) to empowerment level (with action skills, LOC, and intention to act). And each stage has certain knowledge and attitude characteristics (Akintunde 2017).

The **Value Belief Norm** (VBN) Theory was proposed by Stern (2000). This causal chain of variables, grouped into categories of values, beliefs, and norms, influences whether a person is likely to adopt some kind of pro-environmental behaviour. It was suggested that the VBN cluster of variables was stronger in predicting behavioural indicators than other theories, even when other theories were taken in combination (Heimlich and Ardoin 2008).

The **Planned Behaviour Theory**, proposed by Ajzen (2002), considered the intention to act and objective situational factors (such as economic constraints, social pressures, and technical opportunities) as direct determinants of pro-environmental behaviour (Kollmuss and Agyeman 2002). The intention was to summarize the interplay of cognitive variables, which include knowledge of action strategies and action skills; and variables relating to personality, that are the LOC, attitudes and personal responsibility (Ajzen 2002; Akintunde 2017).

6 Factors Influencing Environmental Behaviour

Almost 40 years ago, Heberlein argued that environmental attitudes are based on values and composed of beliefs and affect the environment as an object (Heberlein 1981). He concluded that environmental attitudes are fundamentally important, widely discussed and frequently attempted to measure—but poorly understood. Hines et al. (1987) reported direct relationship between attitude and behaviour. Environmental attitudes have two components, an emotional aspect involving feelings, and a cognitive aspect, which involves factual knowledge and beliefs (Heberlein 1981). Affective influences on environmental concern and behaviour have been widely studied during recent decades (Kals et al. 1999).

In environmental psychology, the physical environment may be perceived exclusively in terms of emotions and associations (Strumse 2007). Emotions play a significant role in the process of reasoning. It seems that people may rely on emotion as much as reason when resolving deciding moral dilemmas (Klarreich 2001). The recognition of our emotional bond with nature is the basis for building respect for the integrity of all life forms on this planet (Taylor 1981). Today, ecopsychology focuses on the connections between the ecological crisis and our dysfunctional relationship with nature (Jijimon Alakkalam 2014). The challenge is to combine "the sensitivity of therapists, the expertise of ecologists, and the ethical energy of environmental activists" (Brown 1995: xvi).

Hines et al. (1987) also identified two types of attitudes: attitudes toward the environment generally, and attitudes toward taking environmental action (e.g. recycling, conserving energy, petitioning); both types were related to environmental behaviour. However, the power of attitudes to influence behaviour declines when other variables are considered (De Young 2000). Dotzour et al. (2002) found that while there is no linear relationship between knowledge and attitudes or between attitudes and behaviours, people must generally possess the proper knowledge and attitudes to take pro-environmental action. This suggests that it is not enough to simply provide people with declarative and procedural information. In order for people to take pro-environmental action, they must also believe that they possess sufficient knowledge and skills to be competent in the action they take (perceived competence) and that their actions are proper (Dotzour et al. 2002).

Attitudinal causes have the greatest predictive value for environmental behaviours that are not strongly constrained by context or personal capabilities. For behaviours that require high costs or efforts, or for those that are difficult, contextual factors and personal capabilities are likely to account for more of the variance (Stern 2000). In addition to understanding the nature of environmental problems, in order to assist people with making the best environmental choices, one needs to know what behaviours can resolve the problems and how to perform these behaviours (knowledge of action strategies) (De Young 1993).

The LOC is a variable that describes how a person feels to be reinforced for performing a certain action. The LOC has two components; an internal LOC implies that the person believes that he or she has the ability to bring about change or impact through his or her own individual action, while an external LOC means that an individual feels that change brought by his or her action is affected only by chance or by a greater external force. Internal LOC leads to behaviour change, since people who are more confident in the effects of their actions are more likely to act, whereas an external orientation tends to discourage it. This variable is most effectively influenced by repetition of behaviours that result in success, but it can be improved through learning about action skills. Development of an internal LOC in students to reinforce for acting in responsible ways is of great importance. Education does not seem to have a direct impact on LOC, but rather it motivates the person to perform certain actions and the success in performing these actions will reinforce an internal LOC. The knowledge of action strategies is also identified as an important variable (Hungerford and Volk 1990).

South Korean researchers found that LOC and attitude are more important than knowledge and personal responsibility when determining the effects on intention to act. Specifically, the core variable "for improving the intention to act for responsible environmental behaviour is internal LOC", which indicates that "the internal LOC should be stimulated to change or to improve the intentions to act" (Hwang et al. 2000, p. 24).

Intent is another of the many factors affecting or predicting behaviour (Stern 2000). Hungerford and Volk (1990) identify intention as a number of variables acting in combination, including knowledge, skills and personality factors. The stronger the intent, the more likely the behaviour, especially when the person acts in a supportive environment. A participatory environment is one that matches the skills and interests of the individual and allows them to interact either cognitively or physically (Dotzour et al. 2002).

A synthesized model of ERB is shown in Fig. 13.3, where cognitive factors include mental capacity, knowledge, beliefs, and worldview; whereas situational factors include social, economical, technological, political, and institutional factors.

Environmentally beneficial actions are often influenced by personal habit or routine, or may also come from "non-environmental concerns, such as desire to save money, confirm a sense of personal competence, or spending more time for social relationships. … [T]he role of environmentalist predispositions vary greatly with

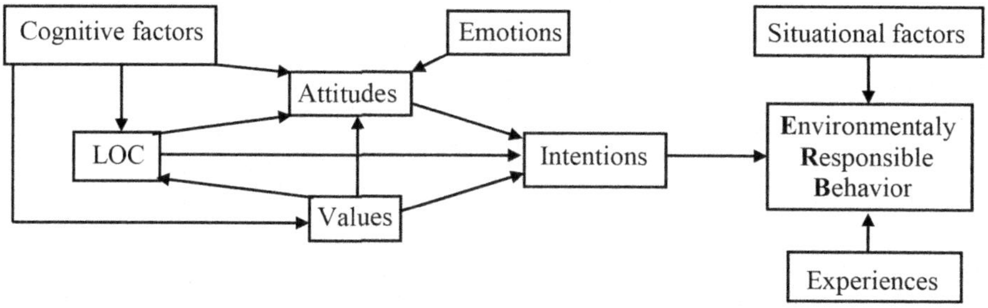

Fig. 13.3 A proposed model for the effects on responsible environmental behaviour (ERB)

the behaviour, the capabilities of the actor, and the context" (Stern 2000, p. 415) and "the causal factors may interact" (Stern 2000, p. 422).

7 Employing Environmental Virtue Ethics in Sustainability Education

As Orr noted, our environmental crisis is not fundamentally one of technology but one of mind, will and spirit (Orr 1992, 1994, 2000). Proper actions require the cultivation of proper virtues (York and Becker 2012). Dealing with values and moral issues is integral part of teaching at all levels of education.

From a cultural-historical point of view, moral development is inherently social. Students have to learn how to reason about morality, and morality is also considered to be a cultural practice in which students must learn and experience how to participate. Classroom discussion is an essential element in this integrated model. Students are encouraged to engage in discussions and to form critical questions concerning the roots and causes, as well as strategies to environmental issues. These relevant moral and ethical questions include, for example: What kinds of responsibilities do we have to wild and domestic animals and even plant species? How do we consider, for example, animal welfare? How does the recognition of rapid, local or global level environmental change challenge our traditional understandings of our obligations? Did the dominant Western and other ideas and worldviews contribute to the escalation of environmental problems—or are these the sources of the problems? What does sustainability mean, and what can an individual do to live a sustainable life? And what are the ethical imperatives for science and society? The knowledge and understanding gained in these discussions help students develop ethical decision-making skills and attitudes such as a sense of community and responsibility for society (Schuitema et al. 2008).

Environmental ethics is a branch of applied philosophy that analyzes the conceptual foundations of environmental values and societal attitudes, actions, and policies to protect and sustain biodiversity and ecosystems (O'Neill et al. 2012). That is, it studies the moral relationship of human beings to, and also the value and moral status of, the environment and its non-human constituents. Environmental ethics have different forms, as human-centered (anthropocentrism) and nature-centered (biocentrism focuses on living organisms, while ecocentrism tend to include abiotic factors such as rivers and systems that include abiotic elements, such as ecosystems and watersheds) perspectives. Biocentric ethicists believe that humans are a part of society of all living creature, which includes not only animals, but also plants, as we are all Earthlings (Rolston 1999).

Ecocentrist and biocentrist ethicists consider the intrinsic value of natural elements to be significant, and that they cannot be reduced solely to economic value as its instrumental or use values to humans. This attitude of respecting species and ecosystems for their own sake can be considered as a consequence of embracing an ecological worldview, as the result of the understanding of the structure and function of ecosystems and processes. It was proven that those people who prioritise intrinsic values have been shown to exhibit higher levels of ecological attitudes and behaviours (Jordan and Kristjánsson 2017), and this orientation usually results in commitments to the protection of species and ecosystems (O'Neill et al. 2012).

Indigenous peoples have developed specific systems of relations with nature, which involved ecological ethics through many centuries of traditional use of natural resources. In the societies of Evenks and Evens, for example, moral imperatives are shaping human subsistence, worldviews, land use, and ritual practices (Sirina 2008). However, their traditional reindeer herding has been reduced in many places due to the use of pasture lands for mining (Brandišauskas 2018).

Rolston (1994) describes environmental values at different organizational levels (organisms, species, ecosystems, Earth, nature). The principles of environmental ethics rests on the principle that there is an ethical relationship between human beings and the natural environment. Human beings are a part of the environment and so are the other living creatures, and plants and animals are an integral part of our lives.

Virtue ethics has a long classical tradition, as Aristotle claimed, virtues are the key determinants of human flourishing. Human excellence encompasses intellectual, moral, and physical characteristics with associated virtues, which will ultimately yield flourishing (Carson 1999, p. 94, cited in Treanor 2014, p. 28). Virtues are character traits (Sandler 2013) and they have multiple components including perception, emotions, and reasoning, which altogether constitute motivation for, and the habit of, virtuous actions (Judson 2015; Jordan and Kristjánsson 2017).

Reflections on character and virtue are prominent in the work of early, influential environmental thinkers. For example, voluntary simplicity (temperance) is central to Henry David Thoreau, whereas Aldo Leopold emphasizes and cultivates love and respect towards the environment. Rachel Carson, like Leopold, believes that "cultivating virtue is central to appreciating the value and beauty of the natural world." For her, "wonder is a preeminent environmental virtue, since wonder and humility are wholesome emotions" (cited in Sandler 2013, p. 1666).

Biophilia is a character trait whose target and affective content is caring for non-human lives. It is an important consideration in governing our relationship with animals and plants, and so it is of particular significance in achieving sustainability goals, and it helps to have rewarding interactions with non-human lives. Biophilia contributes to human flourishing in two ways; by protecting ecosystem services and by fostering connections with the biosphere that nurture the human spirit. To foster biophilia individuals and communities must rely on reliable scientific knowledge about ecology, biodiversity loss, and human environmental impacts. Biophilia, as a virtue, appreciates the complex interactions of organisms in their various environments. All virtues have a cognitive dimension, and part of the cognitive dimension of biophilia is the acquisition of knowledge about our relation to the natural world, as well as about the ecological interrelations of non-human life (Clowney 2013). Biophilia is, however, usually hard for modern people to acquire completely because it often goes against the trends of the consumerist and industrial societies (Clowney 2013).

The essay of Elizabeth Anscombe entitled Modern Moral Phylosophy, published in 1958, is often credited with beginning the contemporary popularity of virtue ethics. The different aspects of enviromental virtue ethics are discussed by Jordan and Kristjánsson in their recent overview (Jordan and Kristjánsson 2017). Treanor argues relevantly, that "environmental ethics should address not just *what* we should do, but also *who* we should be, describing habits, dispositions, beliefs, emotions, and other characteristics"—which is, in essence, character education (Treanor 2014, p. 160).

In the interpretation of Frasz (1993), "an environmental virtue refers to a mean between two vices, qualities the possession of which will partially enable a person to lead environmentally good life" (Frasz 1993, p. 259). Frasz argues, that unlike traditional ethical theories, which focus on the acts of the agent, virtue ethics focuses specifically "on the nature of the agent", that is, "on the virtues of character", and that "the holding of these virtues is intrinsically valuable" (Frasz 1993, p. 260). Environmental ethicists "place ample emphasis on long-term patterns and interconnected relations of the ecosystems. According to such a position, a moral agent who *habitually* looks to the long-term effects of actions will more likely make environmentally sound judgments". The goal of environmental virtue ethics is "to foster *new habits* of thought and action in the moral agent—not just to get the immediate decision made right, but to reorient all actions henceforth in terms of a holistic, ecologically based way of thinking" (Frasz 1993, p. 260). Such an ethic requires much input from various disciplines, pointing to its transdisciplinarity nature.

In contrast to traditional environmental ethics, which "focuses on doing actions because they produce consequences that are…environmentally good" (preferably for the long-term), or because "they are derived from some standard reflecting what is intrinsically good", in environmental virtue ethics long-term consequences are complemented by an immediate good, the environmentally virtuous, which is considered "intrinsically valuable." Therefore, "the basic motivation in environmentally informed virtue ethics is to perform environmentally virtuous actions for their own sake" (Frasz 1993, p. 261). This is a possible critique for environmental virtue ethics because it focuses on the performance of actions for their own sake, rather than on the product or consequences of those particular actions, "leading to a charge that virtue ethics is concerned more with style than substance" (Frasz 1993, p. 263). In other words, virtue ethics is not so much what you actually do, but rather with acting well. As Frasz also argues, "in the long run being virtuous is always the best policy", producing the best and sustainable outcome (Frasz 1993, p. 264).

One of the main challenges for environmental virtue ethics is that we are living in pluralistic societies with many confronting interests, in which there is often no "broad crosscultural consensus concerning what is environmentally right or good" (Frasz 1993, p. 264). This makes finding the best environmental solution difficult. However, it must be stressed that environmental virtue ethics often advocates new *LOCal* (or 'Glocal') solutions inspired by *global* awareness, and societal changes may be stemming from smaller-scale transformations. Those personal changes resulting in environmental friendly lifestyle must be inspired by education—however, the fundamental motivations and attitudes must come from within (Frasz 2001). Joanna Macy (1996) uses the word 'practice', which is a traditional synonym for the cultivation of virtue, by which she means "fortifying the mind and schooling its attitudes" (Macy 1996, p. 503).

Sandler (2005) argues that "environmental virtues are the proper dispositions or character traits for human beings to have regarding their interactions and relationships with the environment. The environmentally virtuous person is disposed to respond—both emotionally and through action—to the environment" (Sandler 2005, p. 3) with inanimate and living creatures. Because there is no particular way to acquire proper environmental virtue ethics, Sandler (2005) defined four regular strategies for discussing the relations between virtues and the environment, that is, for specifying environmental virtues. These are listed and characterized in Table 13.1.

These four approaches to the specification of environmental virtue are not mutually exclusive, but may work in concert, and one can find support from all of them. "Collectively they provide a rich variety of resources for thinking about the substance of environmental virtue" (Sandler 2005, p. 6.). As late András Sütő, a renowned Hungarian writer wrote wisely, good examples must be lived and kept alive—not preached and explained (Sütő 2008).

It is quite common that two (or more) values and two (or more) virtues are confronting. For example, when trees or even larger areas of forest are clear-cut in order to have a field for installing photovoltatic panel arrays. In this case the virtue of protecting life (the trees) or habitat for biodiversity come in to collision with the intention of establishing a more sustainable renewable energy source for human consumption. In order to find a mutually satisfying solution for this problem, a multi-faceted process of consultation with all of the stakeholders is needed.

Table 13.1 Four strategies for specifying environmental virtues (based on Sandler 2005)

Strategies	Extensionism	Benefit to agent	Human excellence	Role models
Description	Extend the inter-personal virtues applied to human relationships which are normative for a particular range of items, activities, or interactions.	Environmental virtues provide material resources or spiritual benefits serving directly for the agent.	Its possessor becomes a good and prestigious person, contributing to higher levels of social cohesion and enhancement for the biotic community.	Studying the character traits of individuals of environmental excellence by examining their life, work, and characters.
Examples	Temperance, generosity compassion, gratitude.	Clean water and air, aesthetic or recreational goods.	Against habitat decline and biodiversity loss, stewardship for the land.	Local or known heroes recognized for intellectual achievements or practical works.

Narratives play an important role in the proper cultivation of virtue (Treanor 2014) and in teaching environmental virtue ethics, so we also use the narrative approach in our integrated system of sustainability education. As, for example, quoting the tale of the old man (shaman) of an indigenous tribe from the film (Apocalypto) directed by Mel Gibson in 2006: "I saw a hole in the eyes of the Man, deep like a hunger he will never fill. It is what makes him sad and what makes him want. He will go on taking and taking, until one day the World will say, 'I have nothing left to give.'"

Over the past 60 years, interest in virtue ethics has continued to grow. However, as Treanor argues, the utilitarian and deontological approaches have remained the preferential philosophical tools when dealing contemporary environmental ethical problems (Treanor 2014). Therefore, much needed to be done by educational programmes focused in environmental ethics at each level for achieving sustainability goals. Integrating the use of scientific knowledge, ethical principles and argumentation in discussing about moral dilemmas is crucial to solve our current environmental problems.

8 The Pedagogical Method Incorporated into This Integrated Model

The **Kentenich pedagogy** or **Schoenstatt pedagogy**, developed by the educator, psychologist and priest Joseph Kentenich (1885–1968), considers education as the service of life, and, in particular, how to connect (psychological) knowledge with life. The Schoenstatt pedagogy, being deeply natural, seeks to relate fundamentally to the human person and the human condition, and also to assist in forming a new culture with nature. It is focusing on healthy and vibrant interactions between nature and humans, and helps to create a culture of life in its intensive relationships (King 1998). The five "guiding stars" as essential features of this educational approach are: (1) attachment pedagogy, (2) movement (life) pedagogy, (3) covenant pedagogy, (4) pedagogy of trust, and (5) pedagogy of ideals. The "foundation stones" of love and freedom are captured in the "guiding stars" this way: love in the first three points, freedom in the final two (Kentenich 1971a, b, 1972).

Education has fundamental importance in establishing networks of healthy and stable attachments and relationships. Attachments not only involve the mind or will, they specifically engage the heart and soul. A vibrant spectrum of attachments is necessary to fully develop one's capacity to love and engage. An effective **attachment pedagogy** must therefore promote a wide variety of attachments—to persons, non-human creatures, natural and human constructed landscapes, things, ideas and values. The Schoenstatt pedagogy accentuates "education through cultivating attitudes" (Kentenich 1972, p. 79), which means that it tends to stress the spirit behind the forms. In other words, the cultivation of attitudes is connected to fundamental virtues such as faith, hope, love, patience, loyalty, purity, humility, and so on. The time dimension of forming relationships and time for allowing quiet reflection and appreciation is also crucial to the soul to set down roots (King 1998).

The fundamental attitude of the soul is a 'habitus', an innate state of the soul. The fundamental tendency indicates that a soul is drawn into a fundamental attitude. The fundamental tendency grows out of the fundamental attitude of loving nature, and it must be strengthened and upheld faithfully until it may become a fundamental attitude, which motivates actions like pro-environmental behaviour. And the fundamental attitude must be faithfully upheld so that new fundamental tendencies can develop out of it—possibly and hopefully, as the influences of effective education programmes.

Kauffman (2010) claims that the reductionist scientific worldview is—alongside the artificial division between science and the humanities, the lack of a shared worldwide framework of values, and secular humanism—one of the four injuries of our secular modern society. He argues that "reductionism teaches us that, at its base, the real world we live in is a world of fact without values, while, in contrast, our lives are full of value" (Kauffman 2010, pp. 10–13). The reductionist ideology has dominated Western science for the last 400 years. This approach has gained a lot of understandings in different branches of sciences and benefits for society, but the need of the holistic way of consideration has become of increasing importance during recent decades (Kauffman 2010). As Capra (1982) argues, rational thinking is linear, focused, and analytic. It belongs to the realm of the intellect, whose function is to discriminate, measure, and categorize. Thus rational knowledge tends to be fragmented. Ecological systems, contrary, are in a dynamic balance based on cycles and fluctuations, that is, they are nonlinear by nature (Capra 1982). Joseph Kentenich stressed the importance of organic thinking in terms of respecting life and also its living context of attachments and relationships. Hence, it is an integrative way of thinking; integrating nature and knowledge, head and heart, life and ideas. It unites the aspects of life that relate to one another (Kentenich 1971a, b; King 1998).

Movement or **Life Pedagogy** considers education a dynamic process of growth, development and life. For Joseph Kentenich (1971b), 'only life begets life'—which is also reflected in current scientific theories, as, for example, in the Biodiversity-related Niches Differentiation Theory (BNDT), arguing that species themselves are the architects of bio-diversity, by proportionally increasing the number of potentially available niches in eco-systems; and the idea of viewing economics, biology and ecology as emergent autocatalytic sets, that is, a self-sustaining network of mutually catalytic entities (Cazzolla Gatti et al. 2017).

Movement pedagogy, therefore, seeks the integration of all of the aspects of life, at both the individual and the community levels. The 'Laws of Life' are essential constituents of movement pedagogy and the effective cultivation of life. Environmental education must always take account of the Laws of Life, that is, life is born from life, life needs material and spiritual resources, life exist in the integrity of structure and function, life has its own dynamics of growing and development, and life is existing in diverse relationships.

Attending to an international Schoenstatt youth meeting in Central America, Kinga Radnai experienced some of the worldview of the LOCal people. There is an established saying of "pura vida" in Costa Rica, she said, which means purity, decent, simple life. Costa Ricans say this in many situations, as a greeting and also for answering questions. By using this, they express their wish to live and experience a pure, decent, and simple life—which is of great significance for realizing a true sustainable way of living (Kinga Radnai, oral contribution, 2019).

The **Covenant Pedagogy** brings the elements of attachment and movement (life) together with concrete personal commitment. To be effective, the process of environmental education must help learners to grow from essential attachments into multi-level relationships. The vulnerability of relationships can very much be experienced. Mutual interpersonal and human-nature relationships may strengthen the dignity of the human person. Covenant pedagogy therefore tries "to lead the person from basic experiences of attachment through personal commitments to the "experience of the covenant"—here the various forms of the environment. Therefore, covenant pedagogy "is a very personal pedagogy, enhancing personal openness and receptivity, sensitivity and sense for values (Kentenich 1971a, pp. 250–251).

The **Pedagogy of Trust** refers to the crucial importance of trust in all forms and levels of education. Because the method of education in Schoenstatt pedagogy seeks to form the whole person from within (that is, through developing and ennobling basic predispositions and desires), appreciating the uniqueness of each person, the trust between educator and educated is crucial. In a pedagogy of trust, the educator cultivates a deep respect for the freedom of the other person(s) entrusted to his or her care. Freedom and trust are also seen in relationship to the cultivation of the spirit of generosity or magnanimity (Kentenich 1971a; King 1998).

The **Pedagogy of Ideals** means to find the appropriate Personal Ideal (PI) fitting best to the personality. The PI enhances spiritualy-inspired motivations and tendencies, that is, the fundamental tenor and temper of the soul. As a psychological definition, it focuses on the intrinsic features of the soul; instead of looking at the ideal from *above*, it looks at it from *within* the person. It also expresses the original natural sensitivity for values and how a person is inspired by his/her personal desires. It is about a special vision that inspires and fills the soul naturally. The PI uses the strong impulses of inclinations, desires, and passions to guide on the way to acquire the fundamental attitude with the impulses of the fundamental tendency. The PI expresses who we are—recognizing our strengths and weaknesses—and what we are called to do as our original mission in our lives.

The twofold function of an ideal, therefore, is that it affirms the unique calling to particular goals (ideal as mission) and it also determines dynamic growth from the present into the future (ideal as identity) (Kentenich 1971a, b). Developing and employing the PI means using our internally-sourced (intrinsic) drivers for constructive purposes, when the PI works like a form of good intention. Then, the PI may be incorporated into the individual actions as conduct, that is, it becomes the core of the personality. A high level of self-recognition is required to find the proper PI for each individual personality, and continuous self-education with firm resolve is also a necessary condition.

The PI has a direct connection to environmental education, for example, in biodiversity conservation, or in taking care of the environment in our everyday lives.

The fundamental orientation of the personality must be reflected in individual decision-making and acts of pro-environmental behaviour (York and Becker 2012). While Joseph Kentenich used the concept of his pedagogy in religious (Christian) contexts, it is equally applicable to effective environmental education programmes and has already been applied in our recent integrated model. The Schoenstatt pedagogy is also already embedded and offers practical pedagogical tools of its own to help promote spiritual and material sustainability.

9 Conclusion

Environmental education has become "a complex and vibrant field of practice and inquiry into the meanings, problems, and potentials of human-environment relationships", as it was well-defined by Brooke Baldauf McBride and her coworkers in their synthesis paper. This diverse theoretical, as well as practical, field of studies is currently of particular importance because of the enormous environmental impacts of our large-scale, effective technologies, and consequential, manifold sustainability problems (Sauvé 2005). Here, we present a complex, integrated model of sustainability education, which has been developing during the last 15 years at the University of Pannonia, Georgikon Faculty, Keszthely, Hungary. This consists of several parts, described above, where the three courses of Human Ecology, Bioethics, and Environmental Education have been taught to over six hundred Hungarian and foreign students. However, this whole system has reached—and, hopefully, influenced—a much broader audience via public lectures. Based on feedback reports and evaluations, this system has proved to be very successful in sustainability education.

Acknowledgement The author expresses his gratitude to the financial support of the grants EFOP-3.6.3-VEKOP-16-2017-00008 and SUMCULA–Erasmus+ Nr. 2017-1-SE01-KA203-034570, respectively, allowing him to study relevant aspects of these complex topics in broad contexts. His thanks go also to his colleague Associate Professor Judit Bódis (PhD) and former students Éva Szabó and Előd Búzás at the University of Pannonia, Georgikon Faculty, Keszthely, Hungary, for their valuable co-operations and contributions to this programme.

References

Adler, J. (2006). Cultivating wilderness: environmentalism and legacies of early Christian asceticism. *Comparative Studies in Society and History*, 48(1), 4–37.

Ajzen, I. (2002). Perceived behavioral control, self-efficacy, locus of control, and the theory of planned behavior. *Journal of Applied Social Psychology*, 32(4), 665–683. https://doi.org/10.1111/j.1559-1816.2002.tb00236.x.

Akintunde, E.A. (2017). Theories and Concepts for Human Behavior in Environmental Preservation. *Fortune Journals*, http://www.fortunejournals.com/articles/theories-and-concepts-for-human-behavior-in-environmental-preservation.html

Athman, J.A. and Monroe, M.C. (2001). Elements of Effective Environmental Education Programs. U.S. Department of Education, Educational Resources Information Center (ERIC), 37–48. http://eric.ed.gov/?id=ED463936.

Babai, D. Molnár, Zs. (2013). Multidimensionality and scale in a landscape ethnoecological partitioning of a mountainous landscape (Gyimes, Eastern Carpathians, Romania). *Journal of Ethnobiology and Ethnomedicine*, 9, 11. https://doi.org/10.1186/1746-4269-9-11.

Boyes, E. and Stanisstreet, M. (2012). Environmental Education for Behaviour Change: Which actions should be targeted? *International Journal of Science Education*, 34(10), 1591–1614, https://doi.org10/1080/09500693.2011.584079.

Brandišauskas, D. (2018). The gluttons of eastern Siberia. Spirits, poachers, and cannibals in Evenki perceptions. http://journals.openedition.org/emscat/3441; https://doi.org/10.4000/emscat.3441

Brown, L.R. (1995). Ecopsychology and the environmental revolution: An environmental forward. *In:* Roszak, T., Gomes, M.E., and Kanner, A.D. (eds.) Ecopsychology: Restoring the Earth, healing the mind. San Francisco, Sierra Club Books (xiii-vi).

Capra, F. (1982). The turning point: Science, society, and the rising culture. Bantam Books, ISBN 0-553-34572-9.

Capra, F. (1997). The web of life: a new scientific understanding of living systems. Anchor Books, New York.

Capra, F. (1999). Ecoliteracy: The challenge for education in the next century. Center for Ecoliteracy, Berkeley, California.

Capra, F. (2011). Systems thinking and the state of the world: Knowing how to connect the dots. *In:* Harding, S. (ed.) Grow small, think beautiful. Ideas for a sustainable world from Schumacher College. Floris Books, Edinburgh. Ch. 6. 72–77. ISBN: 9780863158353.

Carson, Rachel (1998). The sense of wonder. New York: Harper and Row (original publication: 1965).

Carson, R. (1999). Design for nature writing. *In:* Lear L. (ed.) Lost woods: The discovered writings of Rachel Carson. Boston: Beacon, pp. 93–97.

Cazzolla Gatti, R., Hordijk, W. and Kauffman, S. (2017). Biodiversity is autocatalytic. *Ecological Modelling*, 346, 70–76.

Chawla, L. (2006). Research methods to investigate significant life experiences: Review and recommendations. *Environmental Education Research*, 12, 359–374.

Clayton, S. and Opotow, S. (eds.) (2003). Identity and the natural environment: The psychological significance of nature. MIT Press, Cambridge, Mas., ISBN 0-262-53206-9.

Clowney, D. (2013). Biophilia as an environmental virtue. *Journal of Agricultural and Environmental Ethics*, 26, 999–1014, https://doi.org/10.1007/s10806-013-9437-z.

De Young, R. (1993). Changing behavior and making it stick: The conceptualization and management of conservation behavior. *Environment and Behavior*, 25(4), 485–505. https://doi.org/10.1177/0013916593253003.

De Young, R. (2000). Expanding and evaluating motives for environmentally responsible behavior. *Journal of Social Issues*, 56(3), 509–526.

Dotzour, A., Houston, C., Manubay, G., Schulz, K., & Smith, J. C. (2002). Crossing the Bog of Habits (Doctoral dissertation).

Fien, J. and Tilbury, D. (1996). Learning for a sustainable environment: An agenda for teacher education in Asia and the Pacific. UNESCO Principal Regional Office for Asia and the Pacific, Thailand, https://unesdoc.unesco.org/images/0010/001056/105607.pdf

Filho, W.L., Shiel, C. and do Paço, A. (2015). Integrative approaches to environmental sustainability at universities: An overview of challenges and priorities. *Journal of Integrative Environmental Sciences*, 12(1), 1–14.

Frasz, G.B. (1993). Environmental virtue ethics: A new direction for environmental ethics. *Environmental Ethics*, 15(3), 259–274.

Frasz, G.B. (2001). What is environmental virtue ethics that we should be mindful of it? *Philosophy in the Contemporary World*, 8, 5–14.

Heberlein, T.A. (1981). Environmental attitudes. ZfU. 2, 241–270.

Heimlich, J.E. and Ardoin, N.M. (2008). Understanding behaviour to understand behaviour change: A literature review. *Environmental Education Research*, 14, 215–237.

Hines, J.M., Hungerford, H.R. and Tomera, A.N. (1987). Analysis and synthesis of research on responsible environmental behavior: A meta-analysis, *Journal of Environmental Education*, 18(2), 1–8. https://doi.org/10.1080/00958964.1987.9943482

Hungerford, H.R. and Volk, T.L. (1990). Changing learner behavior through environmental education. *Journal of Environmental Education*, 21(3), 8–21. https://doi.org/10.1080/00958964.1990.10753743.

Hwang, Y., Kim, S. and Jeng, J. (2000). Examining the casual relationships among selected antecedents of responsible environmental behavior. *Journal of Environmental Education*, 31(4), 19–25.

Jijimon Alakkalam, J. (2014). A critical appraisal of the uniqueness of ecopsychology as a field of study. *European Journal of Ecopsychology*, 5, 54–63.

Jordan, K. and Kristjánsson, K. (2017). Sustainability, virtue ethics, and the virtue of harmony with nature. *Environmental Education Research*, 23(9), 1205–1229. https://doi.org/10.1080/13504622. 2016.1157681

Judson, G. (2015). Re-imagining sustainability education: Emotional and imaginative engagement in learning. *In:* Selby, D. and Kagawa, F. (eds.) Sustainability frontiers. Critical and transformative voices from the borderlands of sustainability education. Barbara Budrich Publishers, Opladen, Germany. Ch. 11., pp. 205–220. ISBN: 978-3-86649-476-3.

Kals, E., Schumacher, D. and Montada, L. (1999). Emotional affinity toward nature as a motivational basis to protect nature. *Environment and Behavior*, 31(2), 178–202. https://doi.org/10.1177/00139169921972056

Kato, K. (2015). Spiritual and sensory engagement with the more-than-human: An ecohumantities approach to sustainability learning. *In:* Selby, D. and Kagawa, F. (eds.) Sustainability frontiers. Critical and transformative voices from the borderlands of sustainability education. Barbara Budrich Publishers, Opladen, Germany. Ch. 12., pp. 221–237. ISBN: 978-3-86649-476-3.

Kauffman, S.A. (2010). Reinventing the sacred: A new view of science, reason, and religion. Basic Books, New York, Ch. 1. Beyond Reductionism, pp. 10–13. ISBN 978-0-465-00300-6.

Kentenich, J. (1971a). Grundriß einer neuzeitlichen Pädagogik für den katholischen Erzieher. Vorträge der Pädagogischen Tagung 1950, Vallendar-Schönstatt, Schönstatt Verlag. https://www.j-k-i.de/kr-3-de-53/

Kentenich, J. (1971b). Dass neue Menschen werden. Pädagogische Tagung 1951, Schönstatt Verlag, 4-5. Vortrag, pp. 71–84.

Kentenich, J. (1972). Ethos und Ideal in der Erziehung. Vorträge der Jugendpädagogischen Tagung, 28.-31. Mai 1931, Vallendar-Schönstatt, Schönstatt Verlag.

King, H. (1998). Joseph Kentenich—Collected Texts. Free and Wholly Human. Vallendar-Schoenstatt: Patris Verlag, ISBN 3-87620-214-0.

Klarreich, E. (2001). Heart joins head in moral maze. https://www.nature.com/news/2001/ 010920/full/news010920-1.html, DOI:https://doi.org/10.1038/news010920-1

Kollmuss, A. and Agyeman, J. (2002). Mind the gap: Why do people act environmentally and what are the barriers to pro-environmental behavior? *Environmental Education Research*, 8(3), 239–260.

Lakoff, G. (2010). Why it matters how we frame the environment. *Environmental Communication*, 4(1), 77–81. https://doi.org/10.1080/17524030903529749.

Louv, R. (2005). Last child in the woods: Saving our children from nature deficit disorder. Algonquin Books, Chapel Hill.

Macauley, D. (1996). Hannah Arendt and the politics of place: From Earth Alienation to *Oikos*. *In:* Macaulay, D. (ed.) Minding nature. The philosophers of ecology. Guilford Press, New York, Ch. 5, pp. 102–133, p. 127.

Macy, J. (1996). Faith, power, and ecology. *In:* Gottlieb, R.S. (ed.) This sacred Earth: Religion, nature, environment. Routledge, New York.

Mossberg, B. (2005). If trees are us. A relativity theory showing the genius of John Muir's domestic vision of nature for public policy and the national ethos. *In:* Miller, S.M. and Morrison, D. (eds.) John Muir – Family, friends, and adventures. University of New Mexico Press, Albuquerque, Ch. 8, pp. 169–199, p. 193.

Neaman, A., Otto, S. and Vinokur, E. (2018). Toward an integrated approach to environmental and prosocial education. *Sustainability*, 10(3), 583. https://doi.org/10.3390/su10030583

O'Neill, J., Light, A. and Holland, A. (2012). Environmental Ethics. *Nature Education Knowledge*, 3(10), 7.

Orr, D.W. (1992). Ecological literacy: Education and transition to a postmodern world. Suny Press, Albany, New York.

Orr, D.W. (1994). What is education for? *In:* Earth in mind. On education, environment, and the human prospect. Island Press, Washington-Covelo, California, Chapter 1, pp. 7–15.

Orr, D.W. (2000). A sense of wonder. *In:* Barlow, Z. and Crabtree, M. (eds.) Ecoliteracy: Mapping the terrain. Berkeley: Living in the Real World. http://www.ecoliteracy.org/publications/pdf/wonder.pdf

Pisani, F. (2007). Networks as a unifying pattern of life involving different processes at different Levels: An interview with Fritjof Capra. *International Journal of Communication*, 1, 5–25.

Ralph, M. and Stubbs, W. (2014). Integrating environmental sustainability into universities. *High Education*, 67, 71–90. https://doi.org/10.1007/s10734-013-9641-9.

Rolston III, H. (1994). Value in nature and the nature of value. *In:* Attfield, R. and Belsey, A. (eds.) Philosophy and the natural environment. *Royal Institute of Philosophy Supplement*, 36, 13–30.

Rolston, H. (1999). Ethics and the Environment. *In:* Baker, E. and Richardson, M. (eds.) Ethics Applied, Second Edition. New York, Simon and Schuster, Chapter 11, pp. 407–437.

Sandler, R.L. (2005). Introduction: Environmental virtue ethics. *In:* Cafaro, P. and Sandler, R. (eds.) Environmental virtue ethics. Rowman and Littlefield.

Sandler, R.L. (2013). Environmental Virtue Ethics. *In:* LaFollette, H. (ed.) The International Encyclopedia of Ethics. Blackwell, 1665–1674. https://doi.org/10.1002/9781444367072.wbiee090

Sauvé, L. (2005). Currents in environmental education: mapping a complex and evolving pedagogical field. *Canadian Journal of Environmental Education*, 10, 11–37.

Schuitema, J., Ten Dam, G. and Veugelers, W. (2008). Teaching strategies for moral education: A review. *Journal of Curriculum Studies*, 40(1), 69–89. https://doi.org/10.1080/00220270701294210

Sirina, A. (2008). The Living Land: Ecological Ethics of the Evenks and Evens. *Sibirica*, 7(2), 1–22. https://doi.org/10.3167/sib.2008.070202

Sütő, A. (2008). Advent in the Hargita Mountains. *In:* Bertha, Cs. and D.E. Morse (eds.) Silenced Voices. Hungarian Plays from Transylvania. Dublin: Carysfort Press, pp. 21–98.

Staniskis, J.K. and Stasiskiene, Z. (2006). An integrated approach to environmental education and research: a case study. *Clean Techn Environ Policy*, 8, 49–58. https://doi.org/10.1007/s10098-005-0028-1.

Stern, P.C. (2000). Toward a Coherent Theory of Environmentally Significant Behavior. *Journal of Social Issues*, 56(3), 407–424.

Strumse, E. (2007). The ecological self: a psychological perspective on anthropogenic environmental change. *European Journal of Science and Theology*, 3(2), 11–18.

Swan, J.A. (2010). Transpersonal psychology and the ecological conscience. *The Journal of Transpersonal Psychology*, 42(1), 2–25.

Taylor, P.W. (1981). The ethics of respect for nature. *Environmental Ethics*, 3, 197–218.

Thunberg, G. (2019). No one is too small to make a difference. Penguin, Harmondsworth, UK.

Treanor, B. (2014). Emplotting Virtue: A Narrative Approach to Environmental Virtue Ethics. Suny Press, ISBN 978-1-438-45118-3.

UNESCO-UNEP (1976). *The Belgrade Charter.* United Nations Educational, Scientific, and Cultural Organization and United Nations Environment Programme (UNESCO-UNEP) Environmental Education Newsletter, 1(1), 1–2.

UNESCO-UNEP (1978). *The Tbilisi Declaration*. United Nations Educational, Scientific, and Cultural Organization and United Nations Environment Programme (UNESCO-UNEP) Environmental Education Newsletter, 3, 1–8.
Ungerer, M.S. (2015). The effect of environmental education on variables influential to environmentally friendly behavior. MSc thesis, Evergreen State College.
York, T. and Becker, C. (2012). Cultivating an Ethic of Environmental Sustainability: Integrating Insights from Aristotelian Virtue Ethics and Pragmatist Cognitive Development Theory. *Journal of College and Character*, 13(4), 1–12. https://doi.org/10.1515/jcc-2012-1884.

Chapter 14
Education for the SDGs in Teacher Training Using the Method of Unfinished Stories – Rewriting Newspaper Reports

Katalin Hill and Sarolta M. Darvay

1 Introduction

Sustainability education (ESD) in teacher training is very important to help students develop their own sustainability awareness, attitudes and habits in this field and also to help them develop the tools needed for practical ESD. Here teacher training through project work is used to achieve these two goals. This is exemplified by a project task that engaged trainee teachers in writing their own ending for a newspaper article dealing with sustainability issues, as suggested by Rodari's idea of unfinished stories (Rodari 1973). Here, the topic of the story was Cape Town's water supply running out. The project required the trainee teachers to collect information about students' perceptions, ways of thinking and their competencies; to sensitize them to the environmental and social issues, and to present the unfinished story as an educational method.

1.1 Sustainability

In recent decades, the concept of sustainability has changed, expanded and refined. In short and concisely, we could say, "How can we live as a happy person by preserving the natural, social, economic resources and treasures for future

K. Hill (✉)
Faculty of Primary and Pre-school Education, Eötvös Loránd University, Budapest, Hungary
e-mail: hill.katalin@tok.elte.hu

S. M. Darvay
Faculty of Primary and Pre-school Education, Eötvös Loránd University, Budapest, Hungary

Faculty of Education, Department of Biology, J. Selye University, Komarom, Slovakia
e-mail: darvay.sarolta@tok.elte.hu

E. Pénzesné Kónya et al. (eds.), *Environmental Sustainability Education for a Changing World*, https://doi.org/10.1007/978-3-030-66384-1_14

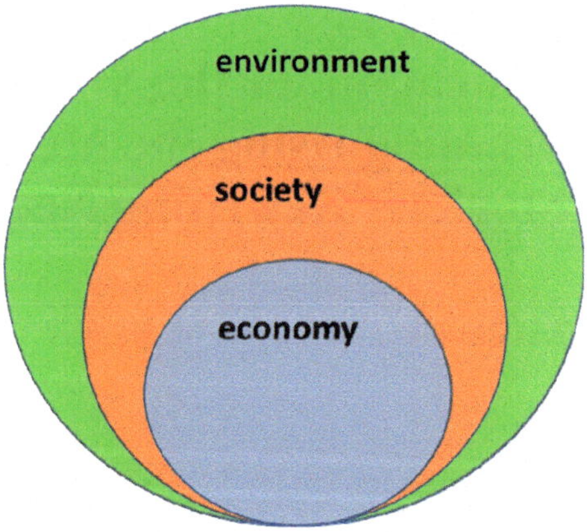

Fig. 14.1 The three main pillars of sustainability

generations". Basically, sustainability consists of three main pillars (environmental, economic and social), so depending on the field of science, the approach to the concept can be very diverse. The environmental pillar includes all ecosystem services and resources (e.g. water, air, soil, wood, etc.). The further two pillars are built upon this foundation (Fig. 14.1). Social resources include cultural heritage, national history, moral standards, relationships between individuals, etc., while economic resources include topics that include financial capital, business and enterprises, technological knowledge, and all aspects of the built environment (Bartus 2013).

1.2 The 17 Sustainable Development Goals

In September 2015, the UN Summit, with the participation of the Heads of State and Government of the world, adopted the Agenda 2030 Declaration. This includes 17 major Sustainable Development Goals (SDGs) and 169 sub-targets. Hungary was the co-chair of the UN Agenda 2030 and greatly contributed to its drafting and creation. Agenda 2030 aims to establish a balance between the three main (environmental, social and economical) pillars. Its 17 SDGs include areas such as poverty, gender and other inequalities, health, sustainable consumption and production, food security, infrastructure, sustainable management of natural resources, climate change, and peaceful and inclusive societies, etc. The 17 SDGs are shown in Fig. 14.2.

Fig. 14.2 The 17 sustainable development goals (SDGs). https://www.bptargetneutral.com/us/how-we-work/choosing-our-projects/the-un-sustainable-development-goals/

1.3 Sustainability Education

1.3.1 UNESCO's Goals for Sustainability Education

UNESCO was the UN agency tasked with the coordination of the 2030 global educational framework. UNESCO defined key competencies, which all students needed to develop in order to achieve these 17 SDGs (UNESCO 2017; Könczey 2017). These are competency in systems thinking, anticipatory competency, normative competency, strategic competency, collaborative competency, critical thinking competency, self-awareness competency, and integrated problem-solving competency. In addition to these key competencies, specific learning objectives for each SDGs are also defined in three main areas: cognitive, socio-emotional and behavioural (Mika and Tóth 2016; Lükő 2017).

1.3.2 Education for Sustainability in Teacher Training

Teachers have a key role to play in ESD, so teacher training must take a special responsibility to prepare future teachers to deliver high quality education for sustainability. In our faculty, we try to integrate sustainability principles with our other main educational goals and the developed of the above-mentioned competencies into all natural science courses. Both in formal, non-formal and informal education, emphasis is placed on personal, experience-based, pedagogy and sensitizing teaching methods.

2 Goals

The aim of this study is to present projects that can be used effectively for teacher training in ESD. First, we briefly refer to the questionnaire survey conducted among our students in the field of sustainability and what has been done to develop an effective training in ESD based on its results. Afterwards, some tried and tested sensitising projects are briefly indicated (Table 14.1) and, finally, one (the 'Unfinished Stories' project) is discussed in detail.

3 Methods

3.1 Questionnaire Survey

Previously, we measured the relationship of our students with sustainability by means of a questionnaire survey (Hill 2015; Hill et al. 2016). In essence, five themes were identified: energy, transport, water, shopping habits and waste pollution. In all areas, students' knowledge, attitudes, habits, social engagement and their willingness to change were explored. In addition, some questions were asked about sociocultural backgrounds that could influence students' responses.

Table 14.1 Short description of some ESD projects

Name of the project	Short description
Making sustainable cosmetics and cleaning products	During the project, we examined first the students' own use of cosmetics and detergents. After searching for environment-friendly and health-friendly solutions, we created some selected recipes together.
Global education by getting to know the lifestyle of people in different parts of the Earth	The aim of this project is to explore the environmental, social, economic inequalities and differences between the living conditions of people living in different areas of the world (altogether 16 countries) (e.g. available drinking water, women's and children's rights, etc.). This encourages global sensitivity and responsibility. Students get closer to their country with family photos (Hill et al. 2016; Anthropolisz 2012).
Composition of land-art	The aim of Land Art is to create a dynamic education based on experience in nature instead of static systems. The materials found in the field inspire the creation of visual communications between man and nature. Photographs of captured spectacles as well as photographed Land Art works can be the starting point for further practice and analysis (Bernáth 2007).
Eco walk	The goal of eco walking is to educate students in a sustainable way. In terms of its educational organization, it is most likely to be a study walk and a study trip, in the area of the school, within a few hours of walking. The walk involves different stations, where students can get acquainted with all three major pillars of sustainability (environmental, economic and social) through playful tasks. Students become more sensitive to long-term sustainable lifestyles in informal circumstances and become familiar with the potential of their immediate environment for sustainability (Dőry and Hill 2016).
Rewriting real stories	Rodari, the famous Italian fairy tale writer, often wrote different endings to his tales, so the reader could decide which one he would like best to identify with. As an example, students looked at newspaper articles relevant to sustainability (eg, Cape Town's water supply is running out). The newspaper article was interrupted at any point and the student wrote their own original which they then read to the group. This was followed by debate, reflection, interpretation of the lesson and its link to the SDGs.

3.1.1 Aims of the Questionnaire Survey

- To collect information on the relationship of students with sustainability topics. To organize effective education, we needed to see what the basics were for the students and also what was the students' thinking to determine what motivation they might have for adopting a more environmentally friendly lifestyle. These included financial motivations, for example, someone is environmentally friendly because they do not have sufficient financial resources to be otherwise (e.g. no money to buy a car, so go by bike), or because someone is aware of the environmental impact of their actions and act accordingly (e.g. go by bike to not pollute the air).
- To focus students' attention on the effects of their daily routine on the environment
- To use the findings to shape the sustainability part of our teacher training.

3.2 *Sensitizing Projects*

Based on these questionnaires, we concluded that really effective work required sensitizing students to the impacts of their actions (Hill and Darvay 2018). Knowledge alone rarely leads to real change in attitudes or behaviour. For this, it is necessary to have an emotional involvement, personal positive experiences, as a result of which a more sustainable way of life becomes their own internal demand. Taking this into account, we started to develop and implement affective, conative, sensitising projects. The topics and short descriptions of the projects are presented in Table 14.1. The Unfinished Stories project will be presented in more detail later.

3.3 *Common Features of the Projects*

Each of the projects described in Table 1 shares the following common features:

- The current "state" (knowledge, attitudes) of the students has been taken as the starting point.
- Each project helps students to recognize and evaluate their own attitudes and decide if they want any change.
- Each project provides an opportunity to get personal experience and/or build knowledge.
- Each project requires independent research from the students.
- Each project is developed over several weeks, which helps deepening and interiorizing.

3.4 *Unfinished Stories Project*

Gianni Rodari's unfinished tales gave the idea that students would write a new end to an event that occurred and was published in a newspaper (Rodari 1973). Rodari is an Italian poet and writer of children's stories who often wrote 3–4 alternative of endings for a tale (Fried 2004). His aim was to give children the opportunity to choose and to teach them that their decisions could influence the outcome of events.

On this basis, students received a newspaper article on the topic of sustainability (Cape Town's water supply is running out). The newspaper article diverted at any point of their choice so that they could create their own ending. The project was carried out with 70 first-year students in the pre – and primary school teacher training programmes.

3.4.1 Goals of the Project

The goals of this project are in good accordance with UNESCO's goals. They include:

1. Collecting information about students' perceptions, ways of thinking and competences.
2. Sensitization.
   - influence on reason, emotion, behaviour
   - shaping awareness, active, environment friendly attitude
   - realization that, in many cases, we can have an impact on the outcome of events
   - local-global view
3. Getting to know the unfinished story as an educational method.

3.4.2 Project Implementation

We selected an article that deals with the current issue of sustainability issues. Lack and contamination of freshwater is a global problem from the environmental, social and also economic point of view. The lack of clean drinking water is a problem for billions of people. Hungary's fresh water supply is good, so the Hungarian people do not suffer the lack of the water in their everyday lives. For this reason, it is important to sensitize them both to the natural resource problem of the water and to the problems of those people who do suffer from a lack of water.

3.4.3 The Project Plan's Steps

1. Sustainability-related newspaper article – 'Drinking water of Cape Town runs out' is identified.
2. The student participants are invited to interrupt the story at any point and . . .
3. Write a new ending, and then . . .
4. Read their rewritten story to their peers in class.
5. This would be followed by group reflection and discussion and the issues raised would then be . . .
6. Connected to the 17 SDGs – solution

4 Result and Discussion

4.1 Evaluation of the Questionnaire Survey

Looking at the students' knowledge base, we expected better results. In order to form positive attitudes, it is essential to have adequate knowledge about how the ecosystem works and to know the consequences of human actions and their impact on the environment. Here, we should have put more emphasis on knowledge during training.

Regarding the student's habits, we got quite varied results, which is not surprising considering that these students may come from very different backgrounds (type of settlement, type of school, parents' pattern, background material etc.).

Overall, we concluded that, in many cases, the students were simply not aware of their impact on the environment (for example, 75% of them underestimated per capita water consumption in Hungary). Considering this, it is very important to know the impact of our lifestyle on the environment and to raise our awareness. In terms of willingness, according to their own admissions, students would sacrifice to make their habits more sustainable. One of our main goals is to increase our students' willingness to adopt more pro-environmental habits based on internal motivation.

4.2 Evaluation of the Unfinished Stories Project

As a first step of evaluation, we examined what kind of strategies were chosen by students to write a new end for the story, ie to solve the water shortage. How many percent of students will find out a solution which solves the problem with conscious water conservation and with a reduction of water consumption and what percent would seek a solution that avoided the need to limit water consumption. Figure 14.3 shows that the strategies of 75% of student's involve maintaining current levels of water consumption while only 25% of the students sought strategy to decrease water consumption. This result is not surprising if we consider that the principle of consumer society is to increase consumption and that decreasing consumption is a clear loss of value.

By analyzing the endings of the stories, we tried to categorize the possible solutions invented by the students. Figure 14.4 shows the percentage distribution of the solutions students suggest in the categories we have created. This is followed by some typical examples of each solution category as written by the students.

4.2.1 Technological Innovation

- People have joined together to build a water desalinization plant.

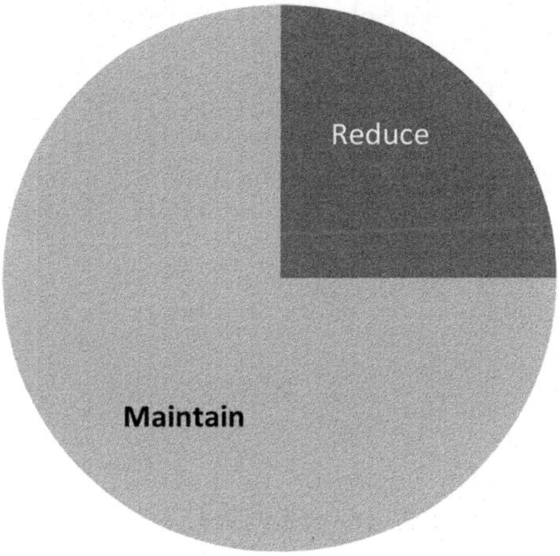

Fig. 14.3 Students' strategy: 75% of them would maintain and 25% of them would decrease water consumption

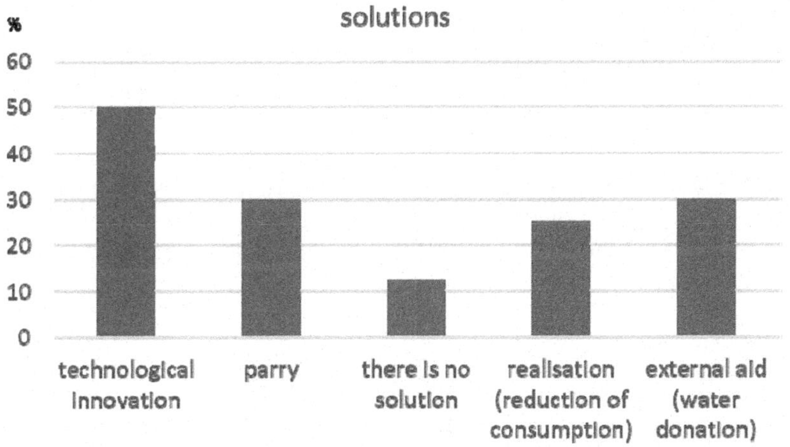

Fig. 14.4 Categories of solutions chosen by the students

- Importation of fresh-water from the melting glaciers of the Antarctic – 'so solving two problems at the same time'.

4.2.2 Parry, Denial

- People do not believe that water can really run out, anti-government strikes are organized.
- Nature will solve the problem, smaller people will evolve so less water is needed.

4.2.3 No Solution

- The city is abandoned because it has run out of drinking water. A National Geographic film has been shot to provide lessons for other people.
- The rich people emigrate, so the poor people have enough water for their needs.

4.2.4 Realisation

- People reduce water consumption, but in return, the government ensures minimal water.

4.2.5 External Aid (Water Donation)

- Those who lived run out of water, but receive help from the UN.
- They pray to God.
- A World Conference was held in which other nations donate water.

We also examined whether students used any kind of collaboration to solve the water supply problem. This could be anything from international to local cooperation. Figure 14.5 shows that half of the students did not find a collaborative solution. The other students' collaborative solutions can be found on the right side of the figure.

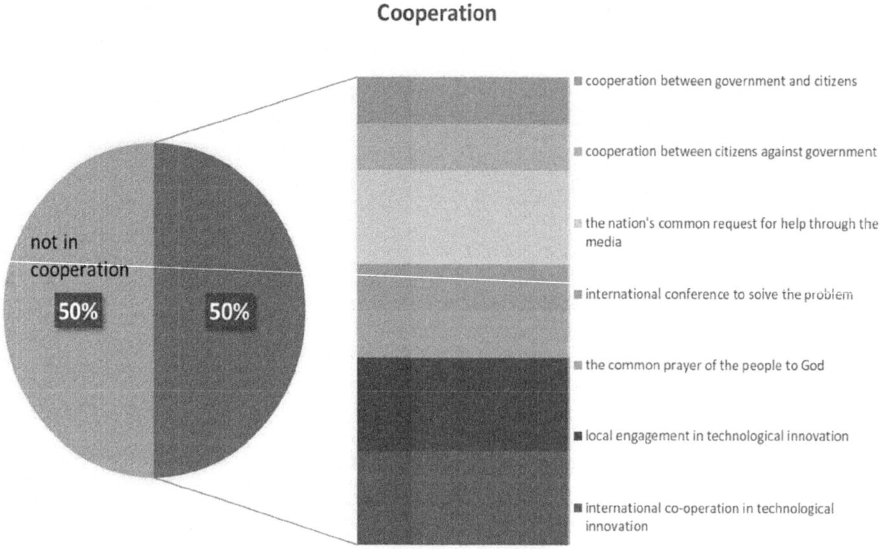

Fig. 14.5 Half of the students tried to solve the water supply problem without cooperation. The other half of the students' collaborative solutions can be found on the right side of the figure

Figure 14.5 contains important information on how young people, future teachers, think about collaboration and cooperation. Real progress in the cardinal issues and challenges of sustainability can only be achieved through cooperation and collaboration, so it is important that students' social competencies also develop.

5 Conclusion

A questionnaire survey was used to examine student's knowledge of and attitudes toward sustainability. The results are of value to the reshaping of sustainability education in teacher training. During the discussion, the opinion of many students was that the survey itself was educative. This sensitizing project was suitable and effective for ESD and for inspiring of attitudinal change. Based on the experience with such activities, students were very active and enthusiastic about the projects. It is important to understand that change is more effective if the student realizes and feels the need to change habits for themselves. This internal change is a prerequisite for embedding the values of education for sustainability in future teachers. It is important to develop a holistic approach to student learning, and these projects have proven very effective in this respect.

References

Anthropolis Assosiation (2012). *Így élnek mások – Felfedező projekt a világ körül.* (The way others live – Exploration projects around the world). Budapest, Anthropolis Association.

Bartus, G. (ed.) (2013). *Nemzeti Fenntartható Fejlődési Keretstratégia*, (National sustainable development framework strategy). Budapest, Nemzeti Fenntartható Fejlődési Tanács, 14–34. http://nfft.hu/nemzeti-fenntarthato-fejlodesi-keretstrategia/

Bernáth, J.A. (2007). *Környezeti nevelés a művészet eszközeivel* (Environmental education with the tools of art (Land Art)), Új Pedagógia Szemle, 3–4 http://folyoiratok.ofi.hu/uj-pedagogiai-szemle/kornyezeti-neveles-a-muveszet-eszkozeivel-land-art

Dőry, J. and Hill, K. (2016). Ökoséta a fenntarthatóságra nevelés gyakorlatában (The Eco-walk in the practice of education for sustainability). *In:* Vitályos Gábor Áron, M Pintér Tibor (Edsszerk.) *Mester és tanítvány II.* Budapest, pp. 50–62.

Fried, I. (2004). *Modern olasz irodalom és színház: problémák és művek.* (Modern Italian literature and theater: problems and works). Bölcsész Konzorcium. Budapest. p. 233.

Hill, K. (2015). A fenntartható életmód vizsgálata és fejlesztésének lehetőségei a tanítóképzésben (Study of sustainable lifestyle and opportunities to develop it in teacher training). *In:* Rajnai Zoltán, Fregan Beatrix, Marosné Kuna Zsuzsanna, Ozsváth Judit (szerk.)*Tanulmánykötet a 6.* Báthory-Brassai nemzetközi konferencia előadásaiból, pp. 617–627.

Hill, K. and Darvay, S. (2018). Fenntarthatóságra nevelési projektek alkalmazása a tanítóképzésben (Use of sustainability education projects in teacher training). *Hazai és külföldi modellek a projektoktatásban Nemzetközi Tudományos Konferencia* Óbudai Egyetem, Rejtő Sándor Könnyűipari és Környezetmérnöki Kar p. 2 ISBN 978-963-449-074-6

Hill, K., Darvay, S. and Zsoffay, K. (2016). Fenntarthatóságra nevelés a tanítóképzésben a hallgatók globális szemléletének alakításáért. *In:* Rajnai, Z., Fregán, B., Marosné Kuna, Zs. (ed.) *Tanulmánykötet a 7.* BBK előadásaiból. Budapest, pp. 383–391.

Könczey, R. (2017, szerk). *Fenntartható fejlődési célok oktatása* (Education for Sustainable Development Goals). UNESCO, EKE OFI letölthető innen:http://ofi.hu/node/181079

Lükő, I. (2017). *Oktatás és fenntarthatóság az ENSZ Fenntartható Fejlődési Célok (SDG 2016-2030) rendszere alapján. Holisztikus, rendszerszemléletű és multidiszciplináris megközelítések* (Education and sustainability based on the system of UN Sustainable Development Goals (SDG 2016-2030). Holistic, system-based and multidisciplinary approaches). EDU 7. évfolyam 3. Szám. http://epa.oszk.hu/02900/02984/00016/pdf/EPA02984_edu_2017_3_007-031.pdf

Mika, J. and Tóth, B. (2016). *Az ENSZ Fenntartható Fejlődési Célok (2016-2030) környezeti vonatkozásai* (Environmental Aspects of the United Nations Sustainable Development Goals (2016-2030)). Héra Évkönyv, 549-569.http://hera.org.hu/wp-content/uploads/2014/02/HERA_evkonyv_kiegeszites20170615.pdf

Rodari, G. (1973). Grammatico della Fantasia: Introduzione all'arte di inventare storie. Piccola Biblioteca Einandi, Torino.

UNESCO (2017). *Education for Sustainable Development Goals*. UNESCO, Paris.

Index

E. Pénzesné Kónya et al. (eds.), *Environmental Sustainability Education for a Changing World*, https://doi.org/10.1007/978-3-030-66384-1

Printed by Printforce, the Netherlands